SATELLITE

SATELLITE
INNOVATION IN ORBIT

DOUG MILLARD

REAKTION BOOKS
*Published in association with
the Science Museum, London*

To Mum and Dad who encouraged me then;
Stephanie, Edward and Daniel who encourage me now

Published by Reaktion Books Ltd
Unit 32, Waterside
44–48 Wharf Road
London N1 7UX, UK

www.reaktionbooks.co.uk

In association with the Science Museum
Exhibition Road
London SW7 2DD, UK

www.sciencemuseum.org.uk

First published 2017

Printed and bound in China by 1010 Printing International Ltd

A catalogue record for this book is available from the British Library

ISBN 978 1 78023 659 9

CONTENTS

v-2 rocket on launch pad, Operation Backfire, Cuxhaven, Germany, 1945. Operation Backfire was a British post-Second World War operation designed to evaluate the v-2 rocket system.

INTRODUCTION

I t is a truism, and a cliché, that we have gazed at and wondered about the heavens since time immemorial. Many an exhibition or even book on space starts with such a statement. The ancient civilizations of Greece, Mesopotamia, India, China and Central America are all likely to be cited as centres of cosmic observation and investigation. In modern times Jules Verne's voyages to the Moon in his literary marriages of fantasy and rationalism are likely to feature too. His contemporary of the nineteenth century, Giovanni Schiaparelli, showed us drawings of Mars that hinted at intelligent life on the red planet. We thought afresh of how to reach these places in space. How would we get to Mars? How could we break the bonds of gravity and reach other worlds?

Some did the maths. They calculated how space flight could be achieved, how fast they would need to go to get into space and by using what type of vehicle. The first step would be to make it into Earth's orbit. You would use a rocket, in stages that fell away as each exhausted its fuel supply. Your rocket would grow lighter as each load detached, and accelerate on to a velocity of 8 km per second. Then your speed would be balancing Earth's pull of gravity and round and round you would go, falling perpetually around Earth as a new moon; a new satellite.

From there – a staging post in orbit – you could depart to the planets and beyond. The rocket would be huge and use exotic, energy-rich propellants; the gunpowder and fireworks that had been around for centuries had merely showed the way. Within decades of such predictions large space rockets were with us, not aiming for space but raining death and destruction onto tens of thousands. The 10-metre-high v-2 rocket of the Second World War was aimed at London but touched space as it climbed ever higher before falling on to its civilian target. Hot war gave way to cold war, and a geopolitical impetus for launching satellites grew as a means of demonstrating the technological superiority of the launching state. They would also be able to conduct scientific investigations, relay television signals around the world, and – especially so – gather intelligence on one's enemy. The first was Sputnik, launched by the Soviets and opening an age of space in which we continue to dwell . . . except we no longer appreciate it. At first satellites were headlines, prestige that dazzled. Now they are invisible, yet vital; ubiquitous yet mysterious.

The satellite is a reticent technology. Our modern world depends on them just as much as the computer, the automobile or the aircraft. We use it to farm, to carry out financial transactions, for moving around the globe, for generating energy, for monitoring the weather and climate, for security and scores of other functions. The satellite is essential.

The story of the satellite is remarkable. Its diffidence cloaks an astonishing history of imagination, experiment and ingenuity. Its history traces the formative years and events of the tumultuous twentieth century. This book unveils the satellite, once trumpeted as the future present, anew. What forms has it taken? How did the forms evolve and why? Who required them? What exactly do satellites do and how? And what larger picture are satellites part of? The satellite was a first step on a road to the stars. We have walked on the Moon, but Mars – that catalyst to the imaginations of the early space thinkers – remains for the future. We are content to stay mainly in Earth's orbit, peering across the cosmos through space observatories, or sending satellite emissaries to other planets and their moons. The pioneering satellite Sputnik and its successors help us on Earth and extend our place in space but pose questions, too, of how far we choose to extend that existence.

Chapter One explores the history of the satellite as a concept of the mind; of how thinkers well before Sputnik, or indeed space rockets, envisaged an object that could travel endlessly around our planet (and others). It taps into those minds that worked out how to reach space and thence the planets; for these theorists, satellites would act as the staging posts in Earth's orbit for interplanetary voyages. Isaac Newton developed the necessary mathematics of the motions involved; two centuries later the Russian schoolteacher Konstantin Tsiolkovsky calculated the speed and means needed to take that first step into Earth's orbit using powerful new rockets. Come the twentieth century and such thinking was invigorated in a period of frantic technological development, via revolution and warfare, when the rocket, no longer just a firework and indiscriminate battlefield weapon that scared the horses and burned down cities, was given enormous power and precision. The German v-2 rocket of the Second World War terrorized and killed thousands, yet pointed the way into space in a manner that the visionaries had long dreamt.

Chapter Two dwells on the post-war years when the satellite began to take shape. The recent hostilities had seen science used as never before to extend and accelerate the prosecution of war. New technologies had emerged within huge state-sponsored research and development programmes. Rockets far more powerful than the v-2 would now be able to reach around the world with nuclear warheads; still more powerful rockets could be used by the military to launch artificial satellites. There was a growing cultural perception of satellites and space flight as an inevitable future. The gap between the science fiction of the cinema and the science fact of the laboratory seemed to be closing. The dormant imaginations of the designers working on the missile programmes, their aspirations kindled decades earlier by the likes of Verne and Tsiolkovsky, were now stirred, and slowly they drew their plans for reaching space, the Moon, Mars and beyond. The future, it was assumed by many in the post-war world, would be American; its affluence and technological might was surely a guarantor for it being the first nation to reach space. This was not to be. The first artificial satellite of the Earth was launched by the Soviet Union in 1957, while the U.S. floundered on the launch pad.

Chapter Three examines the frantic early years of the space age. Reaction to Sputnik was intense, particularly so in the U.S., where Soviet satellites were now perceived as a military and political threat. The U.S. reinvented its muddled satellite programme with new agencies geared to putting the nation firmly in orbit and then on to the Moon. In the USSR a record of space firsts was maintained, and extended to putting the first manned satellites into orbit. Both nations were now fighting a proxy war in orbit, with space a political tool for the competing geopolitical ideologies. For others, satellites became the epitome of a space-age cultural and consumerist trend that marked the arrival of the future.

In the closing period of the Second World War, a young English serviceman wrote of a means by which television, the first public service being little more than a decade old, might unite the world with transmissions beamed via satellite. Twenty years on and this dream of Arthur C. Clarke's materialized with the first satellites launched into the eponymous Clarke orbit. Chapter Four shows how satellite technologies developed by the military elite, primarily in the USA, were rapidly transferred to serve the wishes and needs of humanity as a whole. Satellites could now reach millions, enabling real-time communications around the globe. They were becoming also the technological exemplars of new elites, commercial operators and media giants. The reach of the satellite was spreading, but the military still set the pace in terms of their development. In 1991 the U.S.-led armed forces of the first Gulf War were relying, for the first time, on a sophisticated array of satellite services, including communications, meteorological, reconnaissance and positional. This last capability, first demonstrated in the early years of the space age, had now been refined to an unprecedented degree (in order to provide locations at sea for nuclear submarines, with accuracy to a few metres) and found wide use by all of the armed services. The Global Positioning System, or GPS as it came to be known, was made available to civilian users, so contributing to a developing reliance by societies on all forms of satellite.

Chapter Five looks at how a satellite works, how it is launched, what orbits are used and why. Satellites are, in principle, straightforward objects that need only to survive launch, the space environment and then to actually work. In reality they are constrained by the weight penalty they bring to their launching rocket's performance. They cannot exceed the maximum lifting capability of their rocket. They must be robust yet kept as lightweight as possible. They must be capable of working faultlessly, untended, for years on end in the most hostile of environments. Their payloads are the instruments or systems required for them to carry out their specific task: looking at Earth or out into space; relaying communications signals; providing split-second timing for terrestrial systems and services, location and navigation. And with Sputnik followed by thousands more this chapter considers what happens to the rising population of satellites that, once at the end of their working lives, clutter more and more of Earth's space.

Chapter Six, the final chapter, considers how satellites we have launched into Earth's orbit or the orbits of other planets have helped in our understanding of the universe and our place within it. What is out there? What does it look like and how does it work? The satellite has been turned to for help in answering these questions. Of Earth they now provide previously unimaginable insights of the natural systems that drive our planet and its place in the cosmos. But the sensors of satellites have peered further out into space than ever before, returning dramatic data and imagery of the universe, its structures and exotic behaviours.

Satellites have revealed worlds orbiting other stars and have been sent to the planets of our own solar system, none more so than to Mars, the landscape of which satellites have mapped in more detail

Overleaf
A pair of small CubeSat satellites shortly after deployment from the International Space Station, 2014.

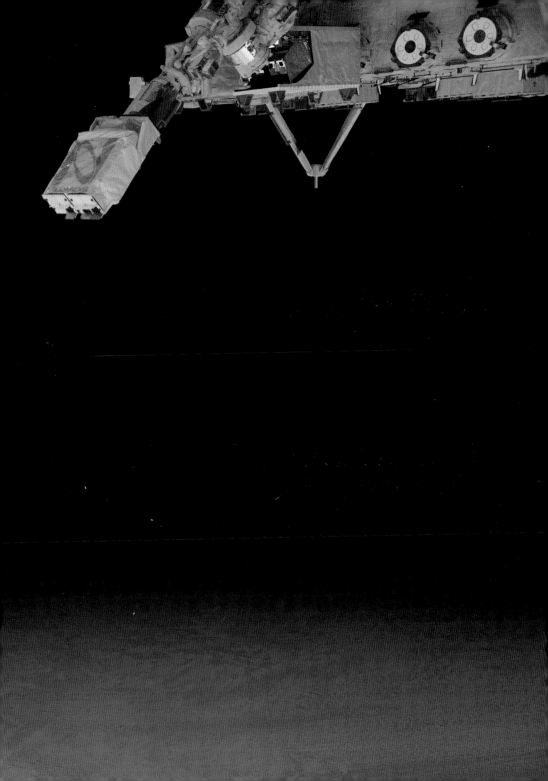

than the ocean depths of our own planet. For centuries the red planet has drawn the gaze and stoked the imagination. H. G. Wells, in his seminal science fiction story *The War of the Worlds* (1898), wrote how 'across the gulf of space ... intellects vast and cool and unsympathetic, regarded this earth with envious eyes, and slowly and surely drew their plans against us.' His allegory of nineteenth-century colonialism substituted hungry Martians for the missionaries and armies of empire. Today Earth continues to look upon Mars with intellects hopefully inquisitive and benign rather than cool and unsympathetic. Those that launched Earth's space age with its first artificial satellites were looking to Mars and beyond. Slowly and surely humanity, with satellites extending its reach into space, is perhaps drawing its plans for getting there. Yet the way remains hugely expensive. We still reach space with the same types of rocket used to launch Sputnik, each discarded after one mission. When we can get into orbit routinely, with reusable rockets, then all manner of far cheaper satellites will be possible, and with that a new space age will dawn.

1
SATELLITES OF THE MIND

An artificial satellite of the Earth is a difficult concept to conjure in the mind: a body launched into space, at a velocity so great – at least 8 km per second – that it circles the Earth continuously. There is an illustration dating from the eighteenth century depicting one of Isaac Newton's thought experiments. He pictured a giant cannon placed at the top of a mountain from which its shot travels so fast and far it has no chance to fall to the ground. It becomes a satellite of the Earth. A satellite will only slow when some force opposes it, usually the gentle but steady resistance of gas molecules spilling away from the upper reaches of our atmosphere. Cultural images of satellites prevail, born of the time after the Second World War when it was believed that science and technology would deliver all. Satellites hurtling endlessly around our Earth like electrons around a nucleus, the launching of rockets and missiles through billows of smoke and flame – these images are quintessentially of the space age. Such imagery conjures up modernity and a break from the past, but the idea of a space satellite had been around in various shapes and forms for centuries, and certainly well before the launching of the very first one – Sputnik – in 1957.

A body in space can be said to be a satellite when captured by the gravity of a larger body, typically a planet, but with the motion of the smaller body (a moon, for example) counteracting the gravitational pull of the larger one. The gravitational pull of each acts on the other, but it is the smaller of the two that circles the larger. Our Moon is

Earth's natural satellite, trapped by Earth's gravity but moving sufficiently quickly to avoid being pulled into Earth's atmosphere.

The etymology of 'satellite' shows very terrestrial origins. The Oxford English Dictionary cites the Latin *satelles*, meaning an attendant or guard, and suggests a reproachful connotation when such a term might have been used: 'implying subserviency or unscrupulousness in the service' offered by the individual. The earliest celestial application of the word was by the astronomer and mathematician Johannes Kepler in 1611 when referring to the moons of Jupiter recently discovered by Galileo (who had called them *Sidera*

Artist unknown, *Johannes Kepler*, 1610, oil on canvas.

medicæa – the Medici Stars – in homage to his patrons). In 1666 the English scientist Robert Hook wrote of 'satelles' when describing his own observations by telescope of Jupiter's moons. And in 1732 the poet Alexander Pope, when musing on humanity and its place in the universe, sought to calm his readers by suggesting we might just as well ask of Jupiter, 'Why Jove's satellites are less than Jove?', implying their subservient relationship to the great planet. Two years later, in his 'Examination of Dr Burnet's Theory of the Earth', John Keill, a Scottish mathematician, included a translation of Pierre-Louis Moreau de Maupertuis' exposition of Cartesian and Newtonian systems, which speaks of the Moon as 'Earth's Secondary or Satellite'.

ISAAC NEWTON

It is perhaps no surprise that it was indeed Isaac Newton, in his tireless quest to rationalize God's creation, who was the first to imagine and set down how an artificial satellite of the Earth would be possible. He was not thinking of a satellite as we would understand it: as a vehicle constructed and launched into space with a practical purpose in mind – relaying telecommunications, surveying weather systems, fixing the locations of cars, trains and even pedestrians on Earth. Newton was

Thomas Barlow,
Isaac Newton,
1863, oil on canvas,
after the original by
Sir Godfrey Kneller,
1689.

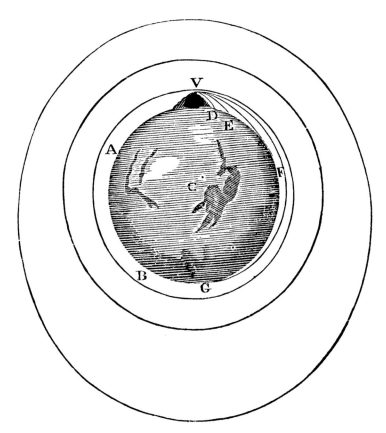

Isaac Newton's thought experiment picturing an ever more powerful cannon on the top of a mountain, accelerating projectiles around the world at orbital velocity. 'Newton's Cannon', from *Principia mathematica* (1687).

merely considering what would happen were a body, at some great height above the surface of the Earth, to be accelerated to a sufficiently high velocity for it to be able to travel all the way around the Earth before falling to the ground – for it to be in a perpetual fall.

In his *Principia mathematica*, published in three books from 1687 to 1726, Newton illustrated his theorizing by imagining a cannon firing a shell from the top of a very high mountain. The projectile would leave the barrel at high velocity and travel a certain distance before eventually falling to the ground. He then pictured a more powerful cannon being used. Its shell would move faster and travel further before once again falling to the ground under the influence of Earth's gravity. Newton then supposed 'either that there is no air about the Earth, or at least that it is endowed with little or no power of resisting' and pursued his model once more with a cannon so large that its ball 'would reach at last quite beyond the circumference of the Earth, and return to the mountain from which it was projected'. He went on: 'when it returns to the mountain, [it] will be no less than it was at first; and, retaining the same velocity, it will describe the same curve over and over, by the same law.'

Newton extended his thinking out into space, reasoning in his *A Treatise of the System of the World*, the third book of the *Principia*, that:

> bodies to be projected in the directions of lines parallel to the horizon from greater heights, as of 5, 10, 100, 1000, or more miles … will describe arcs either concentric with the Earth, or variously eccentric, and go on revolving through the heavens in those trajectories, just as the Planets do in their orbs.

Newton's intuition – supported mathematically – formed part of his treatises on the laws of motion and gravitation that revolutionized people's understanding of how the universe worked. Yet his notes on how a satellite of the Earth could be created remained an intriguing concept which lay dormant, in so far as putting such an orbiting body to practical use, until the nineteenth century.

People had long wondered about reaching and travelling in outer space, but such musings were concerned mostly with reaching other worlds, rather than circling their own. Galileo's pioneering use of the astronomical telescope in 1609 and his published observations of the Moon and discovery of the large moons of Jupiter had demonstrated very clearly the existence of other worlds in the night sky. Here were real destinations to ponder. Francis Godwin (1562–1633), while studying at the University of Oxford, was influenced by the visiting Italian philosopher Giordano Bruno and his then revolutionary beliefs of an infinite cosmos containing distant stars with planets orbiting them. Godwin's story of *The Man in the Moone: Or a Discourse of a Voyage Thither* (1638)

A skein of geese lift the Spanish traveller Domingo Gonsales ever higher, eventually reaching the Moon. 'The Voyage to the World in the Moon', in Francis Godwin, *The Strange Voyage and Adventures of Domingo Gonsales to the World in the Moon* (1768).

employed flights of geese to loft him heavenwards. Cyrano de Bergerac (1619–1655) chose instead, in *The Comical History of the States and Empires of the Worlds of the Moon and the Sun* (1687), to harness himself to phials of morning dew so that the morning sun might draw him towards the Moon. (Eventually he resorted to rocket power!)

THE FANTASTIC AND THE SCIENTIFIC

In the mid-1800s the minds of those imagining journeys through space were informed also by the scientific method. The result was an archaic science fiction or 'scientific romance' in which authors' imaginations were informed increasingly by the swiftly evolving sciences and technologies of the day. Principal among them was Jules Verne, well versed in mechanics, who penned a tale of explorers being blasted on a trajectory to the Moon from the muzzle of a gigantic cannon. His science was fallible, or subject to artistic licence – his intrepid travellers would have been crushed and flattened to the floor of their ballistic shell by its violent acceleration – yet he anticipated elements of NASA's Apollo programme of a century later: he had located his cannon in Florida (NASA's main launch site was built at Cape Canaveral on the Florida Atlantic coast) and the travellers' shell returned from the Moon to splash down in the Pacific Ocean (all Apollo missions did so in the Pacific for retrieval by U.S. Navy vessels).

In one of his less-celebrated stories the main protagonist, Max Bruckmann, describes how he watched as a similar projectile – once again fired from a huge gun but this time as a weapon of war – overshot its target only to carry on and circle the Earth in the manner of Newton's thought experiment. Bruckmann declares in a letter to Professor Schultze, who is menacing Europe and is the owner of the gun, that Schultze's scientific works and inventions are 'odious designs against everything I hold most dear', and, of the shells fired from his super-gun and aimed at 'Ville-France', that

They will fall nowhere . . . A projectile, animated with an initial speed twenty times superior to the actual speed, being ten thousand yards to the second, can never fall! This movement, combined with terrestrial attraction destines it to revolve perpetually round our globe.

The unsophisticated Schultze had failed to do the correct calculations. Nevertheless, Bruckmann goes on to say that this error, 'endowed the planetary world with a new star, and the earth with a second satellite'.

Again, Verne's character – like Newton's mind experiment – is stating how an object accelerated to sufficient speed, altitude and towards the horizon will continue on its way around the Earth ad infinitum. He is not concerned with what function this putative satellite might actually be able to carry out. One of the first to suggest a use for man-made satellites was Edward Everett Hale in his story 'The Brick Moon', first serialized in 1869. Its narrator recounts how, after reading about historical attempts to divine the longitude of a ship while at sea, his brother – known by the nickname 'Q' – described an ingenious plan for launching, over the poles, an artificial moon of the Earth.

The plan was this: If from the surface of the earth, by a gigantic peashooter, you could shoot a pea upward from Greenwich, aimed northward as well as upward; if you drove it so fast and far that when its power of ascent was exhausted, and it began to fall, it should clear the earth, and pass outside the North Pole; if you had given it sufficient power to get it half round the earth without touching, that pea would clear the earth forever. It would continue to rotate above the North Pole, above the Feejee Island place [sic], above the South Pole and Greenwich, forever, with the impulse with which it had first cleared our atmosphere and attraction. If only we could see that pea as it revolved in that convenient orbit, then we could measure the longitude from that, as soon as we knew how high the orbit was, as well as if it were the ring of Saturn.

'But a pea is so small', Q's brother quite reasonably points out. 'Yes', said Q, 'but we must make a large pea.' He suggests further that it be made out of brick, which would offer the necessary strength and would 'stand fire well' during the heating of its passage through the air. This brick moon would have a diameter of 60 m (around 200 ft) and be launched by giant spinning flywheels positioned either side of it. As the moon rolled to them down a massive wooden incline the wheels would whip it up into space and over the Earth. Once orbiting in space it would act as a visible navigational aid for mariners – an elevated Greenwich meridian, traced out periodically by the orbiting brick moon, from which they could chart more easily their passages across the oceans. Hale's invention, it might be said, was a progenitor for the navigational satellites of the twentieth century.

Hale was renowned for the level of detail he included in his stories, instilling them with an authority that suggested the rational. Matters did not go entirely to plan for the heroes of Hale's 'Brick Moon', however. One day the narrator and his friends find the moon gone. It appears that the ground beneath the chute's supporting timbers had subsided, tipping the moon from its securings and allowing it to roll down the gradient to a premature launch. Worse still, some of those constructing and fitting out the moon, and their families – all of whom had moved into the shelter of its quarters during the cold winter months – had presumably been launched too while inside it. There was no news of the moon having crashed to Earth. The whole episode had apparently ended in mysterious tragedy.

Months pass until one day the narrator spots in a recently published astronomical record observation of 'a new asteroid, with an enormous movement in declination'. Weeks later a St Petersburg observatory reports sightings of a new heavenly body moving through the night sky. Encouraged that this may indeed be the lost moon, the narrator and his family set about searching for the object with a telescope, his wife Polly one night suddenly exclaiming, 'It is there! It is there, a clear disk, gibbous shape, and very sharp on the upper edge. Look! Look! As big again as Jupiter!' 'Polly was right! The Brick Moon was found!' the narrator chimes. Better still, so too were its inhabitants: the workers on the sphere and their families, all of whom had been hurled into space during its premature launch, are visible through the telescope. At first they had been assumed lost, the brick moon their celestial tomb. The narrator then notices movement on its surface and realizes the passengers are alive, even appearing to jump up and down in his field of view. Further observation reveals their jumps to be sequential and in patterns; by leaping low and then high from the surface of the moon, they are relaying a succession of Morse code that reports them as being well and contented on their new world. There is a breathable atmosphere, rainfall, and the inhabitants create soil, till it, grow crops and keep hens. There is even a wedding. The happy inhabitants of Hale's Brick Moon had created a satellite space colony and no longer had to concern themselves with the terrestrial travails that would still confront the narrator. Imagining himself in their position, he pictures having no longer to decide 'what Mr Gladstone ought to do with the land tenure in Ireland' and how it would be 'a satisfaction to know that Great Britain is flung off with one rate of movement, Ireland with another, and the Isle of Man with another, into space . . . Victoria would sleep easier, and I am sure Mr Gladstone would.'

The story 'Brick Moon' was not based upon any mathematical calculations, nor included any scientific support for the fantastical journey it put forth. Yet Hale drew heavily for his inspiration from contemporary science, astronomy and politics. It was one of many short stories he wrote alongside historical and biographical works. There was often a strong social ethos to his writing, and he promoted a liberal approach to religion. He was concerned with the betterment of the ordinary citizen, and in this tale space provided the setting for utopia, enabling the formation of a new society in which the people would be free from oppression and hardship. Such sentiments drove others in similar directions. The late nineteenth-century Russian provincial schoolteacher Konstantin Tsiolkovsky, alive to the potential of science and technology for improving the lot of humanity, thought deeply of how space could be reached. He calculated the exact speed that a projectile would have to travel in order to achieve orbital velocity; in so doing this man cemented his name in the pantheon of leading space theoreticians.

KONSTANTIN TSIOLKOVSKY

The young Tsiolkovsky, like many other Russians in the second half of the nineteenth century, had been enjoying a surfeit of reading material – books, magazines, journals, pamphlets and leaflets – as the Russian publishing industries grew. Popular Russian and foreign science and science fiction stories sold particularly well, with the likes of Jules Verne and Camille Flammarion capturing imaginations with tales of space flight and extraterrestrial worlds. Following a bout of scarlet fever at ten years old, Tsiolkovsky read voraciously, the partial deafness that afflicted him having pushed him increasingly into the comfort of his own company. Reading brought him the freedom that the illness had taken away. He had had to leave school and so continued his own education with his father's and brother's books. Later, when staying in Moscow, Tsiolkovsky befriended Nikolai Federov, the librarian of the Rumiantsev Library (now the Russian State Library), who plied him with a far broader range of subjects and titles, especially on history, science, the natural world and philosophy.

Tsiolkovsky became a teacher and lived most of the rest of his life in the town of Kaluga, a couple of hundred kilometres southwest of Moscow. He continued to exercise his mind, thinking about and researching the world around him and humanity's place within it. He achieved a degree of fame with his articles on airships, but by the end of the nineteenth century had started to focus more on space and the possibilities of space flight. He wrote science fiction stories, and in his *Dreams of the Earth and Sky and the Effects of Gravitation* (1896) he outlined the concept for an artificial satellite that could orbit the Earth. Tsiolkovsky thought deeply about how Earth's gravity might be overcome – literally so and as a metaphor for the freeing of the human spirit.

Verne had kindled Tsiolkovsky's curiosity as a boy with visions of space flight, but it was the influence of Federov in Moscow that added a deeper, philosophical rationale for reaching and travelling through space. Federov was the leading proponent of the Russian Cosmist movement, which looked forward to a time when Russian society would lead humanity towards a new state of existence. Federov melded a deep Christian faith with a strong awareness of scientific and technological development. He outlined what he called 'the common task' of uniting all of humankind towards a universal salvation. This would, ultimately, lead even the dead to redemption through resurrection. It was the duty of humanity to pursue science and technology, Federov argued, in order to meet these ends. Once past generations had been revivified, space flight to other worlds would become not only desirable, assisting humankind to achieve a greater harmony with the cosmos, but a necessity as the Earth filled with generations past and present.

Tsiolkovsky's preserved home in Kaluga is filled with the evidence of someone whose inquisitiveness and fascination with the universe was all-encompassing: shelf upon shelf of the books he read and wrote, a study full of scientific instruments – a Wimshurst machine, Leyden jars (vessels that store electric charge), a gyroscope and a camera – and a workshop at the top of the stairs (with childproof trapdoor to help preserve his concentration), where he fashioned his model airships out of corrugated metal. From the workshop there is a door opening out onto the conservatory roof below. Visiting cosmonauts (they all make the journey to Kaluga to pay their respects to this visionary of space) call this Tsiolkovsky's gate to the stars.

Tsiolkovsky set about calculating the velocity needed for a projectile to be able to remain in orbit of the Earth. He reasoned that a rocket would

Konstantin Tsiolkovsky (right) and writer K. Altaisky (left) in the scientist's workshop, 1927. One of Tsiolkovsky's metal airship models stands between them.

be the only realistic technology capable of accelerating the object to such a great speed. Rockets had long been used as weapons of war, albeit rather inaccurate ones, and as fireworks for grand and spectacular displays. Their basic design had changed little over the centuries – a casing, sometimes fashioned from bamboo or paper, sometimes leather (metal in the nineteenth century), with a long stick attached to help stabilize the rocket's flight. The propellant used was black (gun) powder – a mix of ground sulphur, potassium nitrate (saltpetre) and charcoal. It burned unevenly, contributing to the rocket's erratic performance, and despite boosting the missile into the air at great speed, was relatively inefficient.

Tsiolkovsky, however, understood that there was no reason why a sufficiently powerful rocket using a far more efficient, higher-performing

propellant, and then directed more accurately to follow the curva-
ture of the Earth, could not reach orbital velocity. Further, he knew
through Newton's laws of motion that a rocket would be able to travel
on through the vacuum of space; it did not need air or anything else to
'push' against in order to move forward. Indeed, the Earth's atmosphere
would impede its ascent to orbit, especially so in the first few seconds
when it would have to drill through the densest part of the atmosphere
nearest the ground. Newton's Third Law of Motion stated that a force
acting in one direction would impart an equal force in the opposing
direction. A rocket generating a force of thrust with the exhaust gases
of combusting chemicals would impart an equal and opposing force
leading the rocket to move off in the opposite direction. The higher the
velocity of the exhaust gas molecules, the greater the force they would
exert, the greater the opposing force propelling the rocket forward and
the faster the rocket would move.

Knowing the exhaust velocity of black powder to be hopelessly low,
Tsiolkovsky calculated that a rocket using a propellant combination
of the far more energy-rich liquid hydrogen (fuel) and liquid oxygen
(oxidant with which to burn the fuel) would do far better, achieving an
exhaust velocity he calculated to be 8 km per second – sufficient for the
rocket to reach orbital speed. Tsiolkovsky published the formulae for
space flight in 'The Exploration of Cosmic Space by Reaction Devices'
(1903), a treatise which laid out his reasoning and proposals for space
exploration whereby humans, having reached orbit with rockets, could
then live in them (as space stations) and even conduct spacewalks.
Much of Tsiolkovsky's thinking about space was remarkably prescient.
In his sketches for a science fiction film of the 1930s, he described an
airlock that would allow the space travellers to move out and return
safely from within their pressurized rocket cabin or spacecraft into the
vacuum of space. His sketch resembles closely the basic design of the
airlock used by Alexei Leonov during his pioneering spacewalk from
the Voskhod 2 spacecraft in 1965.

IMAGINATION AND ENTHUSIASM

Tsiolkovsky's innovative thinking had little immediate impact – few outside his circle of friends and acquaintances were aware of his work – but in 1924 he republished his writings, stung by a realization that the space enthusiasts Robert H. Goddard in the U.S. and Hermann Oberth in Germany, both of whom were becoming well known for their own rocketry and space thinking and experimenting, were unaware of his work. In 1919 the Smithsonian Institution had published Goddard's theoretical and experimental studies on rocketry and space flight, 'A Method of Reaching Extreme Altitudes'. Goddard went on to launch the world's first liquid-propellant rocket in 1926. His name is now revered in space flight circles, but at the time he gained little widespread recognition. He was methodical and innovative, working up many of the fundamental designs of rocketry that are now routinely employed by the space-launcher business. But he remained aloof, eschewed publicity and was unable to gather the scale of support necessary for the financing of a more ambitious rocket research programme. His

Dr Robert
H. Goddard at
Clark University
in Worcester,
Massachusetts, 1924.

confidence never recovered from
the mocking press reports received
following the publication of 'A
Method', which had included but
a passing mention of how such a
rocket would enable the Moon
to be reached, an ignition of flash
powder signalling its arrival. A few
years after Goddard's publication,
in 1923 Germany's Oberth published his book *Die Rakete zu den
Planetenräumen* (The Rocket into
Planetary Space), and in it he laid
out the principles of rocketry as
applied to space travel.

In Russia, Tsiolkovsky was
contributing to a growing fascination with space flight that, for a
few years in the early 1920s, melded

Hermann Julius
Oberth in 1930.
Oberth worked with
the young Wernher
von Braun and later
was consultant on
Fritz Lang's science-
fiction film *Frau im
Monde* (1929).

powerfully with the utopian ideals unleashed by the revolution of 1917.
Those Russians following a cosmic agenda believed that better worlds
would be built both on Earth and in space; Suprematist artists envi-
sioned existence unencumbered by gravity; Soviet cinema delivered
Yakov Protazanov's *Aelita* in 1924, with its modernistic tale of revolution on Mars; writers Yakov Perelman and Nikolai Rynin promoted
space flight, including the work of Tsiolkovsky, in their publications;
enthusiasts formed societies and in 1927 the world's very first space
exhibit was opened in Moscow as 'The First Universal Exhibition of
Models of Interplanetary Apparatus, Mechanisms, Instruments and
Historical Materials'. The exhibition was truly international, with foreign visionaries afforded as much attention as any, with the works of
Goddard and Oberth exhibited alongside the predictions of H. G.
Wells and Jules Verne. Tsiolkovsky's own situation had veered through

extremes; he and his family experienced terrible poverty, sometimes not knowing where the next meal would come from, and he was arrested and imprisoned shortly after the 1917 revolution for alleged anti-Soviet activity. Yet in his later years he was embraced by the authorities (but not, for many decades, by the Russian scientific elite), broadcasting to the May Day parade in 1935 only months before his death.

A practically minded polymath, Tsiolkovsky experimented and built models and devices in his Kaluga workshop. While airships featured prominently in these workshop creations, rockets did not. The first successful liquid-propelled Soviet rocket (GIRD-09) was not to be launched until 1933. This was done by the Gruppa izucheniya reaktivnogo dvizheniya (GIRD), or Group for the Study of Reactive Motion. The members' goal was to one day build far larger rockets that would be capable of achieving orbit and thence travel on to the planets and beyond. A leading member of GIRD was the aeronautical engineer Sergei Korolev who, a little over twenty years later, would go on to lead the team that built and launched the world's first artificial satellite.

Elsewhere, the German Verein für Raumschiffahrt (VfR – Society for Space Travel, also known as the Space flight Society), had been active since 1927, enthused by the work of Oberth's *Rocket into Planetary Space*. It tested small 'Mirak' and 'Repulsor' rockets – with varying degrees of success – yet the Society's earnest research attracted the interest of the German Army. The high command was keen to experiment with new types of weaponry unrestricted by the terms of the Treaty of Versailles drawn up after the First World War that severely limited the numbers and calibres of guns Germany was permitted. The Army provided the VfR with a disused ammunition dump in Berlin to use as a rocket proving ground. While the quality of the Society's experiments failed to impress – explosions and dud launches were to be expected, but the poor 'quality control' of the research was anathema to the military – the potential of the rocket as a weapon became clear. In 1932, with the Society soon to disband, one of its most able young members – Wernher von Braun – went to work for the German

Army, researching the development of larger and more reliable liquid-propelled rockets. He was a passionate advocate of space flight but knew well that the cost of developing the necessary rockets would be beyond any organization other than the nation's military. Two years later von Braun's team performed the first successful launch of its Aggregate-2 (A-2) rocket, which soared to an altitude of 2 km, higher than any other rocket to date.

In France Robert Esnault-Pelterie derived the rocket equation quite independently of Tsiolkovsky and experimented, too, with rockets (blowing three fingers off his hand when experimenting with a novel propellant). Members of the newly formed British Interplanetary Society were as industrious as any in their advocacy of space flight, but their activities were constrained by the 1875 Explosives Act. This parliamentary legislation prohibited the production and firing of rockets on British soil by anyone other than authorized government organizations.

All of these groups were driven by visions of launching humanity into space and ultimately to other planets, especially to Mars. The red planet had pulled on imaginations around the world ever since the astronomer Giovanni Schiaparelli had published his observations on it in the late 1870s, accompanied by illustrations of its apparent surface features. He named 'seas', which he suggested flooded seasonally when the polar ice caps retreated, and he saw lines and channels of unrelenting straightness and concluded that these geological features were strips of verdant vegetation flourishing either side of flood channels as they filled with water. These networks and lines he called *canali*, which some mistranslated as 'canals', spreading the idea that they were artificial structures, that is, irrigation ditches built by Martians to convey water to the planet's arid areas. The American astronomer Percival Lowell in particular embraced and encouraged this notion and spent much of his life seeking to prove the existence of intelligent life on Mars. H. G. Wells catalysed the allure of Mars still further with the 1897 serialization of his *War of the Worlds*, despite his depiction

GIRD-09, the first Soviet liquid-propellant rocket, 1933, designed by Sergei Korolev and M. K. Tikhonravov and using a mixture of liquid oxygen and benzene paste. Model (scale 1:6).

of the Martians as belligerent aliens set on a hostile invasion of Earth. The German author Kurd Lasswitz wrote also of an advanced Martian civilization in the same year and transfixed imaginations with convincing descriptions of interplanetary trajectories taking humans from Earth to Mars. These were tales of alien worlds and voyages in space that took hold of imaginations in the u.s., Europe and Russia. They invoked enjoyment and wonder but also serious thinking about how these worlds might be reached in the future. In 1924 Mars and Earth were at their closest for more than a century, and this proximity stimulated further interest in both travelling to and communicating with Mars; there were even many reports of strange radio signals received from the red planet. Artificial satellites barely featured in this growing fad for things Martian, but for some travelling to distant worlds would necessitate launching space stations into Earth's orbit first in order to serve as staging posts on their interplanetary journeys.

Hermann Oberth's *Rocket into Planetary Space* included technical descriptions of the systems required for living aboard orbiting space stations. His second book, *The Ways to Spaceflight* (1929), was lighter on the mathematics but heavier on the descriptive, and it carried illustrations showing space stations with spherical living quarters. He also described how the station could be used by scientists and the military.

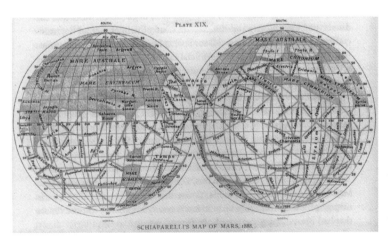

English edition of Schiaparelli's observations of Mars identifying seas (mares) and long interconnecting channels, 1888.

Herbert George
(H. G.) Wells, *c.* 1930.

Herman Potočnik (Hermann Noordung) included in his book of 1928, *Das Problem der Befahrung des Weltraums – der Raketen-Motor* (The Problem of Space Travel – The Rocket Motor), illustrations such as detailed drawings of a circular space station, the design of which echoed down the decades, inspiring others in their work on imaginary and actual space flight. Potočnik's design was dominated by an annular habitat module where the crew would live – a giant ring spinning slowly to generate a centrifugal force and so mimic gravity. The station would be powered by sunlight collected by the massive mirror in the wheel's centre. Two other modules would serve as an astronomical observatory and 'machine' room respectively. The modules would be linked via umbilical connections. It was this type of space station wheel concept that Wernher von Braun, after the Second World War and by that time working for the U.S. Army ballistic missile programme, wrote

Hermann Noordung's
space station habitat
wheel, 1929.

about in an influential series of articles in *Collier's* magazine. With the
artist Chesley Bonestell providing detailed and colourful illustrations,
von Braun enthused Americans with the notion of space flight and the
likelihood of it beginning in just a few years time. A decade later and
the same wheel-in-space design was a central visual feature of Stanley
Kubrick's and Arthur C. Clarke's genre-defining film of encounters
with extraterrestrials, *2001: A Space Odyssey* (1968).

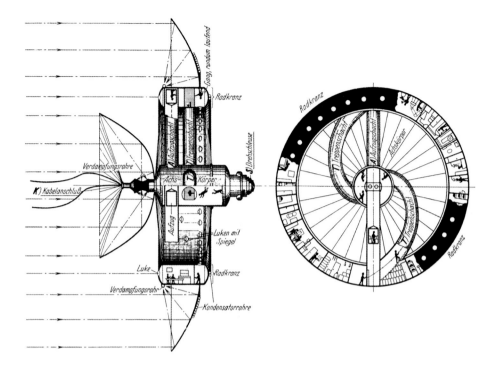

Cutaway drawings
of Noordung's
space station, 1929.
Occupants can be
seen in the annular
habitat area, where
simulated gravity keeps
them from floating.
Radial stairways lead
back to the central
hub, where weightless
conditions prevail.

Wernher von Braun's
1952 space station
concept, depicted
by Chesley Bonestell
in a series of articles
for *Collier's*.

VON BRAUN'S
SPACE STATION
1952
Illustration by Chesley Bonestell

MSFC-71-PD-4000-130

V-2

In the 1930s thoughts of satellites, space stations and space flight had been evaporating quickly as the situation in Europe deteriorated. In Nazi Germany von Braun's team, building on the experience gleaned with the A-2 rocket, had by the end of the decade started construction of the far larger A-4 rocket. This stood more than 10.5 m high and weighed almost 13 tonnes fully fuelled. It would have a range of more than 320 km and carry a 1-tonne warhead. The engine burned alcohol fuel in liquid oxygen. The A-4 was far more powerful than any that had been built before and required a massive research and development programme that cost the equivalent of that spent by the U.S. on the Manhattan Project to build the atomic bomb.

The A-4's propellants were each stored in a separate tank and then pumped to the combustion chamber, where a pyrotechnic device ignited the mix. Early liquid-propelled rockets had relied on gravity to drop the propellants into the combustion chamber. Larger rockets then used gas-pressurized feeds to force greater quantities of propellant into the chamber more quickly. But the A-4 would consume well over 100 kg of propellants per second over sixty seconds to lift off the ground and accelerate to the prescribed velocity and altitude. Von Braun's team developed an innovative propellant turbo pump, modelled on German fire hydrants, to deliver the propellants into the combustion chamber. It was steam-driven from the decomposition of hydrogen peroxide by sodium permanganate. Most of today's space rockets still use equivalent technology, albeit on a far larger scale.

On its flight the A-4 would rise to a height of over 80 km. The engine cut out a few moments before reaching its zenith, leaving it to continue in a ballistic descent onto its target. A far steeper test flight in June 1944 took one rocket to a maximum altitude of almost 180 km. In the 1950s the theoretical aerodynamist Theodore von Kármán galvanized a group of scientists into agreeing to define a boundary between the aeronautical and astronautical realms, henceforward

Space station from the film *2001: A Space Odyssey* (1968), directed by Stanley Kubrick. Its design follows a concept by Wernher von Braun.

known as the Kármán or Karman line. With this nominal boundary between the Earth's atmosphere and outer space set at an altitude of 100 km, the A-4 of June 1944 was in effect the first space rocket. Such a vehicle, despite its military role, brought von Braun's dreams of true space rockets much closer.

The first successful test launch of an A-4 was carried out at the German Army's Peenemünde research centre on the Baltic coast in 1942. Two years later the first operational A-4, now referred to as the V-2 (from the German *Vergeltungswaffen*, or Retaliatory weapons), was launched from a German mobile rocket battery in occupied Holland and struck not far from Paris. The same day another rocket was launched at England and hit west London, killing three people. Over the next few months more than 3,000 V-2s were fired at targets in England, France, Belgium and the Netherlands. Over 1,300 were aimed at London. Thousands of people were killed by the rocket attacks, but many more died as slave labourers building the rockets. The V-2's destructive power was little more effective than the far simpler V-1 flying bomb.

The V-2 was the world's first long-range rocket, and it paved the way for all that followed including the first intercontinental ballistic missiles developed in the 1950s and those in the arsenals of the twenty-first century. Coupled with a nuclear payload, it formed an ultimate weapon system that was near-invulnerable. It ushered in an era of fear, in which nuclear Armageddon dominated the geopolitics of the Cold War. However, it also jolted awake and into the national consciousness those who had been dreaming of space, of launching satellites and space stations into orbit, and of sending spacecraft to other planets. A rocket this large, and which actually touched space during its flight, could be adapted and uprated into a still more powerful vehicle capable of achieving orbit, much as Tsiolkovsky had described. Towards the end of the summer of 1945 several high-ranking German prisoners of war were being interrogated in a requisitioned school building not far from Wimbledon Common in southwest London. Some forty years on I sought out the schoolhouse, attempting to reconcile these infamous prisoners with the familiarity of surburban London. I pictured Wernher von Braun, technical director of the V-2 programme, being driven from leafy Wimbledon, across the bomb-scarred London he had helped create, to daily interrogations at Shell Mex House next to the River Thames. One prisoner, Heinrich Klein, had been a leading designer of artillery and solid-fuelled rockets for the Rheinmetall-Borsig company. Of von Braun he remembered how he kept talking about a far more powerful rocket – the A-9 – one of a sequence of uprated A-4 rockets that would be able to attack Britain and even the U.S. from deep within Germany (A-9/A-10), while an A-12 rocket comprising stages drawn from the other Aggregate designs was proposed to orbit a satellite. As noted in Michael J. Neufeld's *Von Braun: Dreamer of Space, Engineer of War* (2008), 'He went on about the A-9 as the "first intercontinental rocket" but it was only the initial step to a satellite that would orbit at 300 km altitude and 30,000 km per hour velocity.'

2

THE SATELLITES ARE COMING

In 1946 the American Research and Development think tank – Project RAND – published a report called 'Preliminary Design of an Experimental World-circling Spaceship'. This followed discussions and studies by both the U.S. Army and Navy on the feasibility of building an artificial satellite. The forces' interests had been whetted by the preliminary studies carried out at Peenemünde for more powerful Aggregate rockets including one, the A-12, that – according to Wernher von Braun – would have been capable of achieving orbital velocity and therefore of launching an artificial satellite. The RAND study concluded that such a vehicle could be built – a four-stage vehicle with each stage using an alcohol-oxygen motor rocket engine, such that when nearing its maximum altitude, 'the last stage is fired and the vehicle is accelerated so that it becomes a freely revolving satellite'. The costs, however, would be enormous, almost U.S.$150 million (equivalent to a 2016 value of close to U.S.$2 billion).

The report outlined the benefits of such a programme: satellites would help guide the 'high-speed pilotless' missiles predicted of the future, and also monitor the accuracy of their impacts 'and the observation of weather over enemy territory'. Research into cosmic rays and gravitational and magnetic fields was also predicted to be possible from space alongside astronomical studies and investigations of the ionosphere, a portion of the Earth's atmosphere where ionized particles can be used to reflect transmitted radio waves from one part of the Earth to another. Communications could be improved dramatically with the use of satellites as relay stations; if boosted to an orbit of some 40,000 km (25,000 miles) above Earth, so 'their rotational period was the same as that of the earth ... a given relay station could be associated with a given communication terminus on the earth'. In other words, a satellite in this orbit would always be within reach of its designated ground station.

RAND continued its satellite studies over the next few years. It issued a series of reports on the use of satellites for ocean surveillance, reconnaissance and geostationary communications. Yet these proposals struggled to gather any strong support from the U.S. Defense chiefs; the expense and risk of developing novel, unproven technology when there was renewed and massive expenditure on conventional systems in the shadow of the Korean War was too great. Nevertheless, RAND thinking continued alongside a major enquiry into the defence capabilities of the U.S. against a surprise attack by the Soviet Union. Reconnaissance capabilities featured very high in this 'Killian Report' of 1955, named after James R. Killian, the report's chairman, but development of a reconnaissance satellite, the report concluded, was not technically feasible at the time. There were reservations, too, about satellites flying over enemy territory; with the scenario untested, some on the report's investigative panel felt that such an act using military satellites could be interpreted as a provocative action. Mindful of other civilian proposals to develop a satellite, the report recommended therefore that the U.S. develop a small satellite to establish the right of 'freedom of space' – setting the precedent that would enable intelligence agencies to launch larger satellites with impunity in the future. At the time there was no clear understanding of how international law and protocols might apply to outer space. The Killian idea was to force the issue and definition of space as an international domain by launching a small civilian

The stages of a RAND alcohol–oxygen rocket.

PLATE 1
PERSPECTIVE CUTAWAY
OXYGEN ALCOHOL ROCKET PROPOSAL

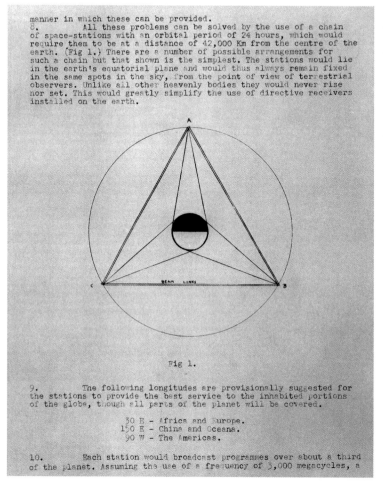

manner in which these can be provided.
8.   All these problems can be solved by the use of a chain
of space-stations with an orbital period of 24 hours, which would
require them to be at a distance of 42,000 Km from the centre of the
earth. (Fig 1.) There are a number of possible arrangements for
such a chain but that shown is the simplest. The stations would lie
in the earth's equatorial plane and would thus always remain fixed
in the same spots in the sky, from the point of view of terrestrial
observers. Unlike all other heavenly bodies they would never rise
nor set. This would greatly simplify the use of directive receivers
installed on the earth.

Fig 1.

9.   The following longitudes are provisionally suggested for
the stations to provide the best service to the inhabited portions
of the globe, though all parts of the planet will be covered.

        30 E - Africa and Europe.
        150 E - China and Oceana.
        90 W - The Americas.

10.   Each station would broadcast programmes over about a third
of the planet. Assuming the use of a frequency of 3,000 megacycles, a

The second page from
Arthur C. Clarke's
*The Space-station –
Its Radio Applications*
(May 1945). The
text and diagrams
are used to describe
'the use of a chain
of space-stations' in
geostationary orbit.

(The Problem of Space Travel – The Rocket Motor), Potočnik
explained how a space station orbiting the Earth above the equator
at an altitude of 36,000 km would appear stationary in the sky were
it visible from Earth. At this altitude its period of orbit would match
that of the Earth's rotation and so, rather than passing over the face of
the Earth, it would remain positioned constantly over the same part.
Clarke further predicted that satellites for world television might be
placed in this orbit within fifty years.

Clarke followed up his letter with an article for the October 1945 edition of *Wireless World* that explained his thinking in more detail, and which the editor chose to title 'Extra-terrestrial Relays'. The satellites would in effect be working as if almost unimaginably tall television masts that, because of their extreme height (many thousands of kilometres rather than a few tens of metres for the real ones scattered over the Earth's surface), would be capable of broadcasting across the entire face of the planet instead of being limited to a horizon on land of a hundred kilometres or so. These satellites, which Clarke also referred to as space stations, would inevitably be large as they would need to accommodate a human crew to carry out maintenance and replace bulky electronic valves when they burned out. Clarke, who went on to achieve great fame as a novelist of science fiction, a science writer and technological prophet, was quite right in predicting the coming of communications satellites. But he had not anticipated the miniaturization of electronics; first the transistor and then the integrated circuit were developed through the course of the 1950s and 1960s – such compact, durable and reliable devices negated any need for Clarke's vast satellites crewed by humans.

Clarke opened the 1995 exhibition at London's Science Museum, appropriately enough, by satellite link from Sri Lanka. The video feed was excellent, showing a chortling Clarke in a University of Colombo lift with its telephone handset pressed to his ear – the only one they could find in the University that would provide the audio link.

BIS SATELLITE CONGRESS

Clarke helped lead the BIS in co-founding in 1950 the International Astronautical Federation, an assembly of similar groups around the world dedicated to the development of space travel for peaceful means. In 1951 the Federation's second congress was held in London, and took the artificial satellite as its theme. The Technical Director of the BIS, Dr Les Shepherd, introduced the assembled to the many uses to which a

Intelstat award, 1990. This was one of Clarke's most prized awards. The inscription reads: 'To Arthur C. Clarke, the godfather of geostationary communications satellites, with appreciation and affection from the father, Harold A. R.' Harold Rosen was the driving force behind the world's first geostationary satellite.

satellite vehicle could be put but made it clear that 'the real value of the orbital vehicle lies in its importance as an essential springboard in the supreme adventure of interplanetary flight. This without doubt must be regarded as the main reason for our interest in the device, all other purposes being of secondary importance.' The sentiment was echoed in a paper, 'The Importance of Satellite Vehicles in Interplanetary Flight', by Wernher von Braun, who, unable to attend in person, wrote that, 'It is one of the main tasks of the satellite vehicle to establish . . . a filling station for interplanetary flights.'

Extracts from the papers delivered at the congress were reproduced in the November 1951 edition of the BIS's journal and a special booklet containing all of the contributions published in 1952 as *The Artificial Satellite*. Another paper outlined the requirements for what was termed a 'Minimum Satellite Vehicle', the simplest and smallest type of satellite that might be built. It influenced the then U.S. scientific attaché to the UK, Fred Singer, who drew up and published his own plans for a 'Minimum Orbital Unmanned Satellite' – MOUSE. It was drum-shaped, about the size of a domestic waste-paper basket and would contain some simple instruments, including a Geiger counter, photoelectric cells and telemetry circuits for communication. For many years Singer's model of MOUSE was displayed in the London Science Museum's 'Exploration of Space' Gallery.

In 1951 Wernher von Braun was working at Redstone Arsenal, Alabama, on the U.S. Army's ballistic missile programme. But he was also, as his 1951 paper makes clear, promoting popular interest in space flight. From 1952 he contributed several articles to a *Collier's* magazine series on exploring space, the colourful illustrations of Chesley Bonestell providing a vivid and convincing depiction of rocket flights to the Moon and other planets. Von Braun was an authoritative and charismatic speaker, and Walt Disney employed him to advise on the design of the space attractions at Disneyland's Tomorrowland theme park and also to help produce, and appear in, a series of television programmes on space. Von Braun was taking a leading role in whetting the U.S. public's

Walt Disney and Wernher von Braun. In the 1950s, von Braun worked with Disney Studios as a technical director, making three films about space exploration for television.

appetite for the inevitability of a coming space age. The future would soon be arriving, and many assumed it would be American.

INTERNATIONAL GEOPHYSICAL YEAR

Elsewhere, an international gathering of scientists had been discussing the implications of the new rocket technologies for the study of the Earth and its atmosphere. In 1950 Lloyd Berkener, Sydney Chapman, S. Fred Singer, Harry Vestine and James Van Allen had met in the U.S.

to discuss ideas for a new year of international scientific collaboration that would exploit these techniques and would be equivalent to the International Polar Years (IPY) of 1882–3 and 1932–3. During the first of these years, investigators from twelve nations had travelled to the Arctic and Antarctic, often experiencing extreme privation in doing so, to observe and record meteorological, geomagnetic and auroral phenomena and to study the currents and movements of the oceans, tides and ice floes. The second of the years, that of 1932–3, built on the earlier research but included more participating countries, and with investigations focused on gauging the effects of atmospheric electrical phenomena at the poles on navigation, wireless communications and telephony. The newly proposed post-war International Geophysical Year (IGY) of research would broaden the scope of investigation still further across a range of physical sciences and would be timed to coincide with the next period of heightened solar activity (1957–8). Emphasis would be given to studies of the upper atmosphere via balloon and the newly developed sounding rockets. Sounding rockets carry scientific instruments to high altitude in order to study space and the upper atmosphere or for photographing the Earth and its surfaces. They follow high, slender parabolic trajectories – ascending steeply before falling back to Earth. Early U.S. sounding rockets employed surplus V-2 missiles as their first stages. The UK Skylark sounding rocket returned the first scientific data during the civilian IGY, although the information gathered, especially on the composition of the upper atmosphere, would be of interest to Britain's military scientists as well, keen to learn as much as possible about the high-altitude environment that ballistic missiles would pass through on the way to their targets.

The year, the organizers suggested, might also see the first satellites launched to assist in these investigations. It was this proposed programme that U.S. military planners subsequently chose to support so as to establish, with a civilian satellite – as outlined in the Killian Report – the principle of free space, paving the way for the launch of

James Van Allen, after whom Earth's radiation belts are named, with an early sounding rocket that would be launched from a balloon (Rockoon), 1955.

intelligence satellites. A resolution calling for development of artificial satellites as part of the IGY's programme was duly adopted by the organizers in October 1954.

Launch of a 'Bumper-WAC' sounding rocket comprising a WAC Corporal missile sitting atop a German-made V-2, 1950.

British Skylark high-altitude sounding rocket, 1960. One of the world's most successful sounding rockets with 441 launched between the first in 1957 and the last in 2005.

SOVIET SATELLITE PLANS

By this time Soviet plans to develop an artificial satellite were themselves moving forward. In May 1954 the Soviet Politburo had instructed
the country's main missile design bureau (OKB-1) to start production
of what would be known as the R-7 intercontinental ballistic missile.
The OKB-1 bureau's chief designer, Sergei Korolev, had long talked with
colleagues about the potential for adapting this missile for launching a
satellite into orbit. He was a remarkable man, a talented engineer but
also a hugely energetic and gifted manager who knew how to get things
done in Soviet society. His mental strength was singular, crucial not
only to his career but in helping to keep him alive when he had been
incarcerated in the Gulag in 1939.

Korolev's deep interest in space had developed while working with
GIRD in the early 1930s. Initially, he had been keen to learn how rockets
might be used to improve the performance of aircraft, boosting their
speed and altitude. He took over the group's leadership after the sudden
death of Fridrikh Tsander, whose rallying call to the membership was
'Onwards to Mars'. Korolev went on to become chief engineer of the Jet
Propulsion Research Institute, formed from the merging of GIRD and
the Army's own rocket research laboratory in 1933. His former GIRD colleague Mikhail Tikhonravov, who had led the development of GIRD-09,
the first modern Soviet rocket, back in 1933, became a keen proponent of
satellite studies. He was working at the NII-4 rocketry research institute
during the late 1940s and early 1950s and, with Korolev's teams, outlined
the possible uses for an artificial satellite. Official reaction to their satellite work was dismissive, but support was more forthcoming from the
powerful USSR Academy of Science. There, Mstislav Keldysh, the leading Soviet mathematician who was playing vital roles in the nuclear and
missile programmes, proved instrumental in gathering support within
the Academy for a scientific satellite programme.

Just a few days after the Politburo's May 1954 ruling on the R-7
missile, Korolev submitted a report prepared by Tikhonravov entitled

Sergei Lyushin (left)
and Konstantin
Artseulov (right)
with Sergei Korolev
and his Koktebel
glider, 1929.

'On the Artificial Satellite of the Earth'. It provided the results of
the research conducted by Tikhonravov's group at NII-4 and also
included an assessment of the supposed satellite work going on in the
United States. Korolev sent the report to the Minister of Armaments,
Dmitriy Ustinov, whom he knew to be sympathetic (unlike many
of his colleagues) to the satellite studies, and three months later the
Soviet Council of Ministers gave it their approval. In April 1955 the
Academy of Sciences created the first Soviet organization devoted to
space flight. It would be chaired by the physicist Leonid Sedov and was
called the 'Interdepartmental Commission for the Coordination and
Control of Work in the Field of Organization and Accomplishment
of Interplanetary Communications'.

In May 1955 in America formal approval for the U.S. IGY satellite
project was first given by the IGY committee and then by the National
Security Council. President Eisenhower approved the project on 27
May. In July the President's press secretary publicly announced that
the U.S. would launch a satellite during the IGY. A few days after that
Leonid Sedov issued a statement at a press conference at the Soviet
Embassy in Copenhagen, saying that 'the realization of the [Soviet]
satellite project can be expected in the near future'. Events were now

Designer Sergei
Korolev (left),
physicist Igor
Kurchatov (centre)
and mathematician
Mstislav Keldysh
(right), 1951.

moving quickly on both sides of the Atlantic. A momentum was building, with interested parties seeing the satellite as a scientific tool, a
military possibility, a unifying entity, a future.

The U.S.'s deliberations on which satellite design to pursue were
protracted. There had been three competing proposals from each of
the Armed Forces: the Naval Research Laboratory's (NRL) 'Vanguard',
the Army's 'Project Orbiter' and the Air Force's 'World Series Satellite'.
Project Orbiter, a small 2-kg sphere, was led by von Braun, who felt its
minimal design and use of existing rockets, derived from V-2 technology, made it the most likely to be chosen. It had considerable support
from the civilian scientific community and, at first, some support from
the NRL. There was resistance elsewhere, however, as the head of the
U.S. ballistic missile programme advised that science satellites were not
important and would divert valuable resources from the missile programme. There was some disquiet, too, about von Braun's involvement
and the launch vehicle's V-2 heritage. Further, some considered that it
would be beneficial for the Navy to develop its own Vanguard rocket,
as it would gain experience of ballistic missile development that at the
time was largely the preserve of the Army and the Air Force. The Navy's
'Vanguard' project was eventually chosen – although the decision was

close – perhaps to assist the Navy in its missile aspirations or indeed to satisfy a perceived need to use a civilian-based satellite programme to set the precedent of 'free space' (at that time, the NRL was still considered to be involved in scientific rather than military affairs).

Back in the USSR the risks of the U.S. attempting a successful satellite launch with a novice rocket were very apparent to the rocket designers. When Premier Khrushchev visited the OKB-1 bureau in February 1956, Korolev assured him on two fronts: that the Soviet satellite work would not interfere with missile development plans and that the satellite itself would be launched successfully because its R-7 rocket was ready (in reality it was only nearly ready: the first successful R-7 test launch was not carried out until summer 1957), whereas the U.S. Vanguard was not. At this time the Soviet satellite design was for a large and sophisticated spacecraft weighing 1.5 tonnes, far bigger and more complex than Vanguard (or, indeed, the Soviet satellite that would

Dmitriy Ustinov as Marshal of the Soviet Union, 1979.

Fridrikh Tsander, a pioneer of rocket construction, 1920.

R-7 'Semyorka' intercontinental ballistic missile design (8K71), 1957.

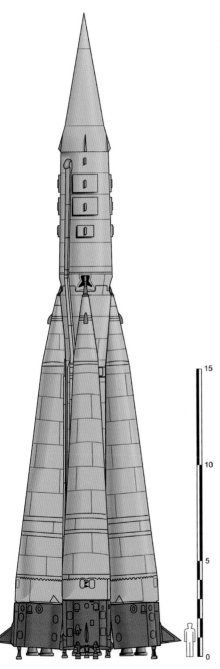

eventually be launched first – Sputnik). It had been given the go-ahead in January with the USSR Council of Ministers issuing decree number 149-88SS, authorizing the launch by an R-7 missile of an artificial satellite termed 'Object D'. Object D was the designation for just one type of possible R-7 payload, other examples (such as Objects A and B) being different designs of nuclear warhead casing.

As the IGY neared, however, Korolev was still facing problems with the R-7 missile itself, which was underperforming and generating insufficient thrust. There were delays, too, in the Object D development schedule, dependent as it was for many of its components on a number of separate organizations and subcontractors. Korolev was worried particularly over reports suggesting that a recent U.S. launch from Cape Canaveral may have been a satellite attempt (it wasn't). He duly asked the Council of Ministers in early 1957 to approve plans for a modified satellite design that could be launched by a slightly less powerful R-7 rocket. This 'simple satellite' would be of a far less ambitious design than Object D. It would weigh little more than 80 kg – one-fifteenth the weight of Object D – and would be designed and built by Korolev's own OKB-1 bureau and just two other supporting organizations, making its construction far easier to supervise. It would also be spherical. This shape would offer the greatest physical strength for minimal

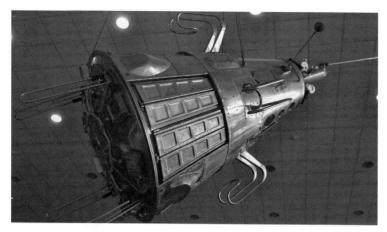

Sputnik 3, launched in 1958 but originally intended as the first satellite to be launched by the Soviet Union when known as Object D.

material outlay and would also optimize the gathering of upper atmos-phere density data as it circled the Earth. Two pairs of trailing stick radio antennae would complete its simple appearance. It is said Korolev insisted the orb be polished to a shine. This satisfied those engineers concerned that the small satellite would overheat in the intense glare from the Sun; anything that maximized its ability to reflect sunlight would help. It has also been said that Korolev equally wished for it to be aesthetically pleasing as mock-ups and replicas would one day be displayed in the museums and galleries commemorating the space age. Sixty years on his prediction was borne out, as confirmed when I inspected both satellite designs at the same bureau's own museum (by this time known as S. P. Korolev Rocket and Space Corporation Energia), prior to their shipment to the Science Museum's Russian space exhibition.

The Council of Ministers gave its approval, and full development work on what would become known as 'Sputnik' ('Fellow Traveller') began. As the year 1957 drew on Korolev's team encountered further difficulties with the R-7; after two failed launch attempts they only managed a first successful test launch on 21 August. Vanguard, too, was suffering problems and significant cost overruns, which were signed off by a very reluctant and sceptical President Eisenhower, not at all convinced of the need to race ahead with a satellite launch.

Sputnik on Show, 1957. A replica of the first Sputnik is seen on display in the Science Pavilion of the Industrial Exposition in Moscow.

Technician working on the soon-to-be-launched Sputnik satellite, 1957.

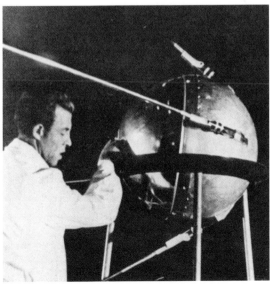

U.S.'s ability to defend itself against Soviet aggression. When Vanguard was chosen as the first U.S. satellite, there was little concern that the Soviets might launch first, despite a series of announcements from the mid-1950s onwards forewarning the world of the Soviet Union's intention to launch a satellite. The possibility was clearly there for those who listened. In the summer of 1957 the Soviets even gave out the frequencies (20 MHz and 40 MHz) their satellite would be transmitting on so that amateur radio stations in the USSR and beyond would be able to pick up the signals. When such signals were duly transmitted from Sputnik in space, the world was stunned. Reaction was intense, a mixture of awe and some concern, too. In the UK the *Manchester Guardian* newspaper asked 'Next Stop Mars?' but added, 'The achievement is immense. It demands a psychological adjustment on our part towards Soviet Society, Soviet capabilities and – perhaps most of all – to the relationship of the world to what is beyond.' France's *Le Figaro* wrote

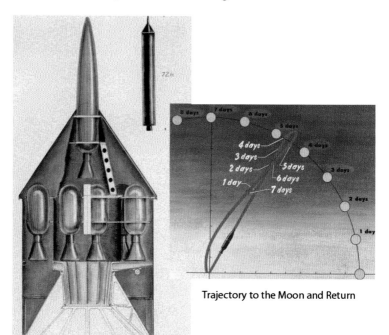

Project Red Socks would have been a U.S. response to Sputnik with a series of nine rocket flights to the Moon. A modified Red Socks was carried out in the Pioneer 4 project in March 1959.

High Speed Stages

Trajectory to the Moon and Return

PROJECT RED SOCKS

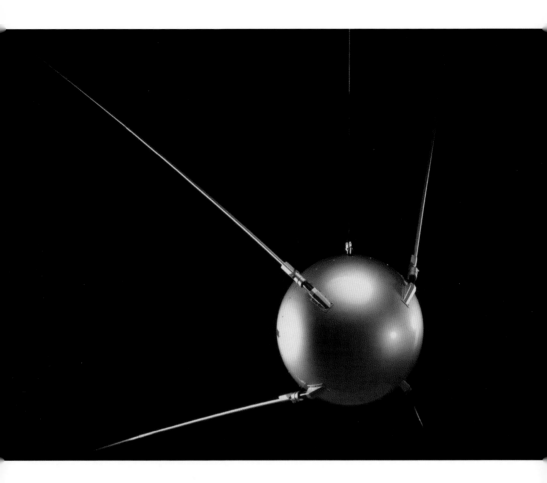

Sputnik 1, 1957
(full-size model).

of the 'disillusion and bitter reflections' of the 'Americans [who] have had little experience with humiliation in the technical domain'.

    Sputnik symbolized the new space age, which was merging with the developing consumer societies either side of the Atlantic and beyond. Space toys packed the shelves and filled the mail-order catalogues. *The Jetsons* zoomed across the airwaves in the U.S., while *Fireball XL5* was launched onto television screens by AP Films in the UK. Architects such as John Lautner moulded the sweeping aerodynamic lines of the futurists into flying-saucer homes. There were Sputnik table lamps and chairs – even a 'Sputnikburger'. Sputnik had become an emblem of

Sputnik curtain by
Vologda lace-makers,
Folk Art Museum,
Moscow, 1963.

A rocket on a launch
pad from the science-
fiction film *I was a
Sputnik of the Sun*, 1959.

modernity, a brand that has even survived into the twenty-first century. It was a name that stuck.

It would be two months before the Vanguard rocket's own engines burst into life and lifted itself and its satellite into the air . . . to a distance of just a few metres before falling back and exploding in a ball of flame and smoke. It was reported that the football-sized satellite could be seen rolling across the launch pad. This very public embarrassment (performed, unlike the Soviet launch, in front of network television cameras) added to what was becoming a major sense of frustration and humiliation in the United States. A month earlier the Soviet Union had trumped the success of Sputnik by launching a second satellite, this time with a dog on board. Sputnik 2 was orbited on 3 November 1957 and Laika, its passenger, became an overnight celebrity. She was the first animal to orbit the Earth, with obvious implications for how long it might be before the first human followed. Of more concern to the U.S. military was the weight of this second Sputnik: although Laika's compartment weighed about half a tonne, the total mass orbited by the Soviets (conjoined compartment and final rocket stage) was well

Laika, the dog the Soviets launched into space.

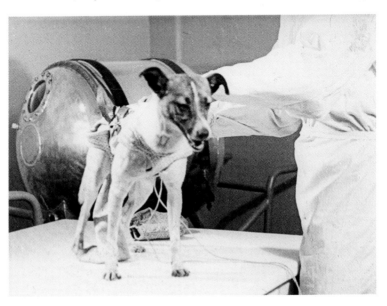

Launch of Jupiter-
C/Explorer 1 at Cape
Canaveral, Florida,
on 1 February 1958.

over 5 tonnes. This underlined how far in advance of the U.S. the Soviet missile programme appeared to be. Launching a satellite of this size was clear proof that the Soviet Union had a missile capable of hitting the U.S. with a thermonuclear warhead.

Finally, and to the relief of many Americans, on 1 February 1958 the U.S. placed its own satellite into orbit. 'Explorer' was a derivative

The Explorer 1
satellite atop its
Jupiter-C launch
vehicle at Cape
Canaveral, 1958.

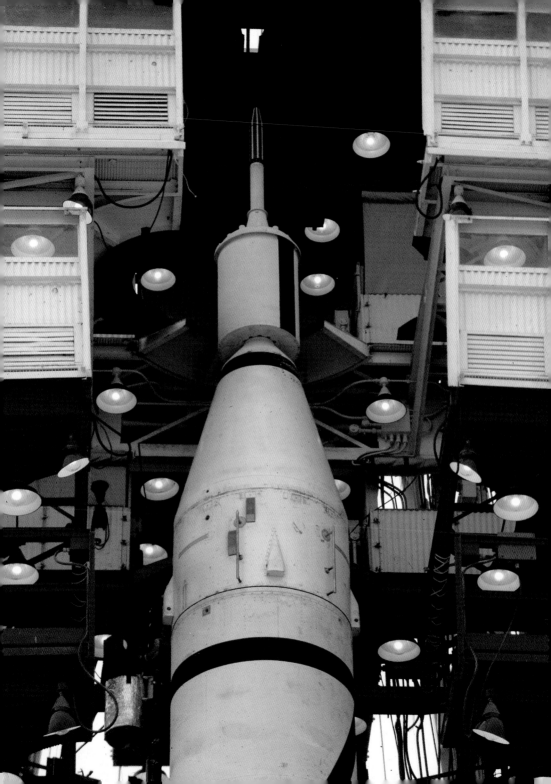

Technicians assembling Explorer 1 on the fourth stage of its Jupiter-C launch vehicle at Launch Complex 26, Cape Canaveral, Florida.

of the original U.S. Army satellite design concept 'Orbiter'. It was not spherical, like Sputnik or the ill-fated Vanguard satellite. It was in effect a small rocket itself, the fourth stage of a Juno 1 launch vehicle: Wernher von Braun's Army team had developed Juno 1 from the Jupiter-C sounding rocket, in turn derived from the Redstone short-range ballistic for which von Braun had drawn heavily on V-2 technologies. Explorer was just over 2 m long and weighed in at 14 kg. Two months later the U.S. finally launched its Vanguard satellite. Chairman of the Council of Ministers, Nikita Khrushchev, took great delight in referring mockingly to the U.S.'s second effort, at 165 mm diameter and weighing just 1.3 kg, as 'that grapefruit satellite'.

The Sputniks continued to fire the world's imagination. The new space age was a participatory event, which ordinary folk could enjoy. On clear nights the spent upper stages of the Sputniks' carrier rockets

William Pickering, James Van Allen and Wernher von Braun holding aloft a model of Explorer 1, 1958.

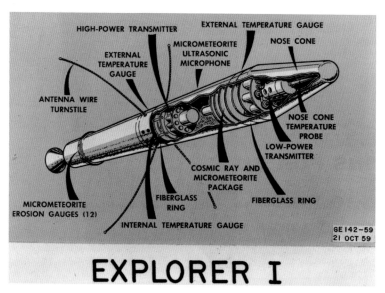

could be seen passing overhead by the naked eye. The University of
Manchester's newly commissioned radio telescope at Jodrell Bank
in Cheshire, England, was able to track the carrier rockets by radar.
The satellites themselves were too small to be followed visually, but
newly commissioned Baker Nunn cameras enabled scientists to capture images of them. These cameras had been the brainchild of Fred
Whipple, Harvard astronomer and director of the Smithsonian
Astrophysical Observatory, who had also played an instrumental role
in the setting up of the IGY. Whipple's astronomical research had
included studies of meteors as they entered the atmosphere, and it was
the experience gained in capturing these events on film that enabled
him to organize the design and construction of the Baker Nunns. Just
two such cameras were stationed in the mainland U.S., with the remain-
der scattered across the world at sites in Spain, Iran, South Africa, India,
Japan, Australia, Hawaii, Peru, Argentina and an island in the Antilles
archipelago off the coast of Venezuela.

  This optical system was supplemented by a radio tracking network
of Minitrack receivers. These had been especially designed by the

Mock-ups of Sputnik
1 and 2 in the Soviet
pavilion of the World
Fair in Brussels, 1958.

Lecturer at the
Moscow Planetarium
explains Sputnik 4's
orbit, 1961.

U.S. Navy to pick up the IGY satellites' radio transmissions, which
international convention had decreed would be broadcast at a frequency of 108 MHz. By using pairs of Minitrack stations, the position
of the orbiting satellite could be calculated by simple trigonometry.
The stations would be located a known distance apart, their separation corresponding to a certain whole number of wavelengths of
the satellite's transmitting signal. The transmissions from a satellite
passing overhead and at a certain angle would, however, reach the
two stations at slightly different times, the difference corresponding
to a certain number of wavelengths. As the satellite passed overhead
the angle would be steep and the wavelength difference small. As it
approached a zenith and then departed, the angle would be shallow
and the corresponding wavelength difference detected by the respective receiving stations greater. Therefore, by noting the wavelength
differences through the Minitrack system, scientists would be able to
calculate the corresponding angle of the satellite in orbit. Minitracks

Tuning into Russia,
10 October 1957.
Jodrell Bank workmen
hammer in stays to
carry cables for the
attempt to track
the Russian satellite.

had been placed down the east and west coasts of the U.S. and South America respectively, with one located in Australia.

Despite the initial international agreement over what radio frequencies IGY satellites would broadcast on, the Soviet Union later made it clear that their satellite would be transmitting at different frequencies, but no adjustments were made to Minitrack. When Sputnik launched, the U.S. was tuned to the wrong frequencies and unable to catch any of the satellite's signals. To compound difficulties, optical tracking was also delayed by the Baker Nunn cameras not being ready in time.

MOONWATCH

Fred Whipple had also organized a civilian satellite-watching service intended to work in tandem with the Baker Nunn cameras and the Minitracks. This would provide a mass data source to help calibrate and hone those readings obtained from the 'official' observations. Such data would be especially important when tracking any satellites with decaying orbits (the early ones were orbited at relatively low altitudes and quickly impeded by the Earth's atmosphere) as they fell towards Earth. With the Baker Nunn and Minitrack problems, these citizen observations turned out to be invaluable.

This combined programme of amateur and professional observation was called 'Moonwatch' and drew heavily from the contemporaneous Skywatch, a civil defence programme in which ordinary Americans

Moonwatch satellite demonstration in Biloxi, Mississippi, *c.* 1957.

could participate by organizing themselves into groups to scour the skies for Soviet aircraft. Moonwatch tapped into this community spirit, with some astronomers noting the range of society it drew from: a Moonwatch cartoon of the time claimed 'Old men shiver, junior high kids romp, local bankers rub elbows with the guy who sweeps out the bank.'

Moonwatch caught the mood of Cold War watchfulness and space-age anticipation. Whipple had publicized it widely in specialist and popular publications. Even Donald Duck was pressed into service to promote the scheme. Specially designed telescopes were distributed to over a hundred groups across the U.S. and more overseas (Japan had a particularly active Moonwatch movement). Moonwatch was a singular collaboration between the professional and the amateur. It tempered the seemingly unstoppable rise of the elite scientist, engineer and technologist – professionals who had honed their skills during the Second

Volunteer Moonwatch satellite trackers in Pretoria, South Africa, c. 1957.

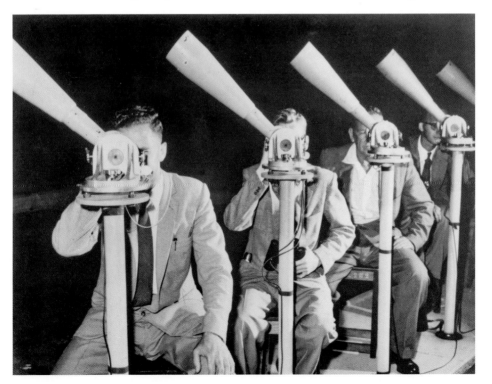

World War and were now burnishing them during the Cold War. Now the ordinary American could be part of this scientific future as well.

The collaboration could be active, extraordinarily so on occasion. As part of a test for Moonwatchers across the U.S., Whipple's team collaborated with the Air Force and Civil Aviation Control authorities to arrange for planes to fly across parts of the country trailing a line and 'satellite'. The 'satellite' was a plumbing plunger fitted with a small battery-powered electric light. Each aircraft would fly at an altitude of 2 km in the evening, the glowing plunger representing a satellite in orbit catching the Sun's rays. The Moonwatchers could use this trailing light as a target to practise their observation skills.

One of the most successful Moonwatchers was Arthur Leonard, a mechanical engineer who enjoyed a passion for astronomy, and whose Moonwatch team usually comprised his wife and son. He was particularly good at finding lost satellites and in 1959, following a one-off reported sighting of the first Vanguard satellite's spent rocket

Boy Scouts participating in Operation Moonwatch, Philippines, c. 1957.

– which had not been seen for a year – he calculated when and where he thought it should appear. It duly did and other groups were then able to corroborate Leonard's sighting using the figures he produced. The now-functioning Baker Nunn cameras were then used to track and photograph the orbiting rocket. Leonard went on to conclude that the satellite's errant rocket had strayed on account of its continuing to accelerate, albeit intermittently, to a higher orbit after its engine should have shut down. This was important new knowledge duly fed back into u.s. rocket design.

THE SATELLITE PREDICTION SERVICE

Another notable Moonwatcher was based in the UK. Russell Eberst mainly worked alone but still produced impressive rates of observation. During one night in 1972, 'he made 127 observations of some seventy-two transits of forty different satellites'. He used one of the advanced Moonwatch telescopes but referred also to the satellite predictions being produced in the UK and made available in an equivalent, if more restrained, programme to Moonwatch. The UK's Satellite Prediction Service had been started in 1957 by the National Almanac Office based at the Royal Greenwich Observatory at Herstmonceux, Sussex. The predictions were distributed by mail to anyone interested, with instructions of how to observe and record the sightings of Sputnik (and its rocket). The observer was encouraged to watch the satellite (with binoculars, if possible) until it passed between two identifiable stars, at which point the observer would start a stopwatch. The respective distance between the two stars would be estimated in tenths so that, for example, the satellite could be observed as being three-tenths from Star A (and therefore seven-tenths from Star B). While watching the satellite the observer should also be listening to the General Post Office's Speaking Clock on the telephone and would stop the stopwatch when the next specific time was read out. This simple method allowed an accurate fix of the satellite's trajectory to be obtained, and the information would then be sent back to the Prediction Service by mail or by telephone. The Royal Aircraft Establishment (RAE) took charge of the Prediction Service in 1958.

The Service's issued predictions were in the form of a map of the northern hemisphere, a transparent overlay containing the track of the satellite in question and a sheet of data. The map would be issued just the once, the overlay every two weeks as the satellite data were adjusted (incorporating the information returned from satellite-watchers). In 1958 sets of predictions and some 3,000 satellite track overlays were issued to almost 200 addresses in the UK and abroad. The service was popular but time-consuming for the RAE personnel, and so transferred to the Radio Research Station (RRS) at Slough, which in 1958 was also designated a World Data Centre for satellites. In the same year Britain's science academy, the Royal Society (RS), which had already played a prominent role in the IGY, sent out a letter of invitation to those interested in watching and recording the passing satellites. To help facilitate this voluntary observing, the RS loaned out binoculars, stopwatches and star atlases, much as the Moonwatch programme provided telescopes.

A central figure in the UK's story of satellite tracking and observation was Desmond King-Hele, a defence scientist based at the RAE. Although interested in the satellites' orbits themselves, he was using their data to learn more of the Earth and its atmosphere. Orbiting satellites were ready-made experiments for finding out about the density of the atmosphere at different altitudes, times of day and year, and also about the strength and nature of high-altitude winds. They would also, naturally enough, be orbiting within the Earth's gravitational field, and their tracks responding to any variations in that field. By following these satellite perturbations, King-Hele would be able to develop revised models for the Earth's gravity field. He and colleagues drew heavily on the public's observations alongside the data returned by suites of high-tech instruments equivalent to those used in the U.S. King-Hele spoke of his work at a witness seminar at the Science Museum in 2002 (and, in private conversation, of what appeared to be a far deeper interest in the life of Erasmus Darwin, grandfather of Charles).

THE U.S. RESPONSE TO SPUTNIK

The Soviet achievements in space, trumpeted by Khrushchev as evidence of communist superiority over the West, triggered a major reassessment of U.S. space, science, military and education policy and the way they were organized. A new Advanced Research Projects Agency (ARPA) was created in January 1958 by President Eisenhower to bring some coherence and direction to the military's hitherto disjointed and competing space projects. Later in the year responsibility for a civilian space programme was invested in a newly formed National Aeronautics and Space Administration (NASA), while the National Defense Education Act set out to raise standards of teaching in science and mathematics.

ARPA moved quickly to meet the challenges facing the U.S. in space. By the end of the year 'Project SCORE' had placed a communications satellite in orbit, although it was essentially just a transmitter attached to the core of its launching rocket. It was an intriguing event on the way to achieving live communication by satellite, as predicted by Arthur C. Clarke and others. SCORE was the first American satellite to relay a radio message from space and was launched by an Atlas rocket (the first time this converted Air Force ballistic missile had been used as a space launch vehicle). The satellite orbited at a relatively low altitude, giving it an orbital period that took it around the globe in just over 100 minutes. It carried a tape recorder with a pre-recorded message

The 1959 NASA seal, commonly known as the 'meatball' logo. The sphere represents a planet, the stars symbolize space and the red chevron signifies aeronautics. A spacecraft orbits around the planet.

that was transmitted down to Earth when the satellite came into view of a receiving ground station. Its Christmas message came from President Eisenhower:

> This is the President of the United States speaking. Through the marvels of scientific advance, my voice is coming to you from a satellite travelling in outer space. My message is a simple one: Through this unique means I convey to you and all mankind America's wish for peace on Earth and goodwill towards men everywhere.

Fears of nuclear war were manifest, however, and came close to being realized with the Cuban Missile Crisis of 1962. The Soviet Union had positioned offensive missiles on Cuba and removed them only after President Kennedy threatened severe consequences if they

Artist's impression of u.s. Vela satellites shortly after separation from their launch vehicle, *c.* 1964.

Assembling a Vela
satellite, *c.* 1963.

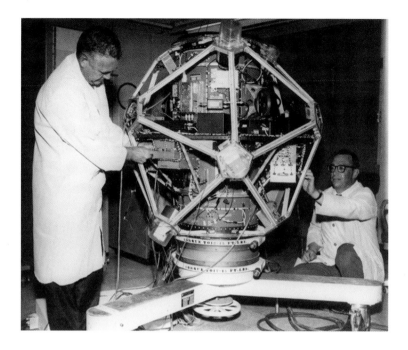

remained. In 1963 the superpowers moved closer in their efforts to
promote a peaceful world. The Test Ban Treaty of 1963 prohibited
its parties from carrying out nuclear tests underwater, at sea, in the
atmosphere or in outer space. In October of that year, soon after the
Treaty came into effect, the U.S. launched two ARPA satellites designed
to detect any ground-, atmosphere- or space-based nuclear explosion.
These were called Vela Hotel 1 and Vela Hotel 2 and were placed in an
orbit 115,000 km out in space, well beyond the Van Allen belts – layers
of charged particles that surround the Earth, trapped by the planet's
magnetic field. *Vela* is Spanish for 'watchman': Vela Hotel moni-
tored atmospheric and space tests; Vela Sierra, ground tests and Vela
Uniform, underground tests. The Van Allen belts, predicted to exist
by scientists, were confirmed by the U.S. Explorer satellite. Readings
had been detected earlier by Sputnik 2 but not corroborated. A second
pair of satellites launched the following year in 1964 detected China's
first nuclear test at Lop Nor. These defence satellites used detectors

that responded to the intense release of gamma rays from a nuclear detonation, but paved the way for their future use in the new field of gamma-ray astronomy.

New ARPA satellite systems were also developed for assisting with navigation. In 1714 the British Parliament had passed the Longitude Act, which provided financial incentives to anyone who could solve the problem of how to determine one's angle of longitude while at sea. The target was to achieve an accuracy of not more than one degree of longitude, or 96 km (60 miles) at the equator. The clockmaker John Harrison surpassed this with his H4 marine chronometer on two occasions, posting an error of just 1.8 km (1 nautical mile) on the first voyage trial to Jamaica and of 16 km (10 miles) on the second. His intricate technology revolutionized navigation in the eighteenth century. Some 250 years later ARPA's Transit satellite system was launched to provide the U.S. Polaris submarine fleet with an accuracy of location at sea of 120 m. This would enable their commanders to calculate an accurate trajectory for their missiles that would lead them to their target. The idea for Transit sprang from the scientists at Johns Hopkins University, Baltimore, who had tracked the Sputniks by measuring the natural change in the satellites' radio transmission frequencies as they passed overhead (also known as the Doppler Effect). By studying these changes they were able to calculate the orbits of the satellites. It occurred to them that the reverse calculation would enable an accurate means of fixing one's location on Earth: a satellite's known orbit would provide the necessary information to calculate a position on Earth. Transit, like the Global Positioning System (GPS) satellites that replaced it in the 1990s, provided data both for military and civilian users.

ARPA needed also to supply the military with good images of cloud cover to assist in the timing of reconnaissance missions. In April 1960 the very first weather satellite, called TIROS (Television Infrared Observation Satellite Program), designed by the Radio Corporation of America (RCA) and managed by ARPA before being handed over to

The TIROS satellite undergoing vibration testing at the Astro-Electronic Products Division of RCA in Princeton, New Jersey, 1960.

The TIROS
meteorological
satellite being fitted
to its spin-up
mechanism, 1961.

the newly created NASA in 1959, was launched. It used two television
cameras to return thousands of scanned images of clouds and weather
systems. The simple utility of this new satellite – delivering real-time
pictures of the world's weather – prompted NASA's development of
a succession of satellites (Nimbus) followed by NOAA's (National
Oceanic and Atmospheric Administration) own series which today
provide a key component of the world's meteorological, climatic and
environmental monitoring systems.

TIROS had been an offshoot of the U.S. Air Force's proposed WS-117 reconnaissance satellite development programme of 1956. The CIA split the unwieldy bundle of WS-117 required technologies into discrete programmes when the Agency took charge of it in 1958. One component (SENTRY) became a research programme on photographic reconnaissance that would employ radio transmission of images. Another (MIDAS) looked at the use of infrared detectors to monitor missile launches. A third (DISCOVERER) worked on a film

recovery satellite design in which film taken by a satellite in polar
orbit would be ejected into a small re-entry capsule that would return
through the atmosphere and be caught in mid-air by a recovery aircraft. Another part of the WS-117 package was looking at the use of
television cameras for reconnaissance, and it was this project that was
redirected into what became TIROS (the resolution and clarity of late
1950s television technology was too poor for military requirements,
but quite adequate for meteorological needs). WS-117, renamed as
Corona by the CIA and eventually operated by the Air Force, was
given the 'cover' name of Discoverer. Early flights brought little success, but in August 1960 Discoverer XIV returned more than 1,000 m
of film of Soviet territory, which provided evidence to dispel the idea
that the Soviet Union had large numbers of intercontinental ballistic

Corona satellite
capsule catching gear
deployed from an Air
Force C-119J aircraft,
c. 1960.

An Air Force JC-130B aircraft practising satellite capsule capture at Edwards Air Force Base, California, 1969.

Arrival of Discoverer XIII capsule, the first successfully recovered, at Andrews Air Force Base, 13 August 1960.

Index camera lens
from a Corona
satellite, made by the
ITEK Corporation,
United States,
1966–72.

missiles (ICBMs). Moreover it showed that the supposed missile gap between the two powers was untrue. A few years later President Lyndon B. Johnson made clear how significant he considered the military satellite programme to be:

> I wouldn't want to be quoted on this but we've spent 35 or 40 billion dollars on the space program. And if nothing else had come out of it except the knowledge we've gained from space photography, it would be worth 10 times what the whole program has cost. Because tonight we know how many missiles the enemy has and, it turned out, our guesses were way off. We were doing things we didn't need to do. We were building things we didn't need to build. We were harboring fears we didn't need to harbor. Because of satellites I *know* how many missiles the enemy has.

Film take-up reel
from Gambit KH-8
surveillance satellite
– the successor to
Corona – made
by Kodak, United
States, 1966–84.

SOVIET ZENIT

The Soviet's own satellite reconnaissance programme was to have been
of a similar design to the early Discoverer, but following Sputnik's
political success priority was given to achieving more space firsts (lunar
and planetary missions). In late 1958 the chief designer Sergei Korolev
merged the reconnaissance satellite project (Object OD-1) with that
being planned for the first human orbital missions (Object OD-2).
From OD-2 Korolev developed the 1K capsule design that would
prototype the fully functional human-rated 3K capsule (2K was the

designation for the photo-reconnaissance capsule design). Later on each was given a new name, with 2K becoming the Vostok-2 (reconnaissance) and 3K the Vostok-3 (manned). In August 1960 the Soviet Union carried out the first space capsule recovery mission. The capsule was of the 1K design and called Korabl-Sputnik 2. It had carried two dogs, Belka and Strelka, on a one-day orbital mission of Earth. Their safe return was widely publicized but with nothing mentioned of the ulterior role of their mission: not only to test recovery of a manned capsule but as a mission of reconnaissance. Two years later (and after Discoverer XIV) the Soviet Union launched its first completely successful reconnaissance satellite mission, which by this time was known as 'Zenit' (zenith), although the secrecy of its mission was maintained by giving it the ubiquitous name of Kosmos, in line with all such military or failed missions. The change to Zenit was necessary following the flight of a Vostok-3KA capsule number 3 on 12 April 1961. Inside was air force pilot and first-ever spaceman Yuri Gagarin. Vostok became synonymous with the human space race that ensued. Zenit went on to become one of the most successful and prolific Soviet and Russian space programmes, with hundreds launched over its thirty-year history.

MANNED SATELLITES

The headlines, however, were to stay with the manned 'satellite' programmes of the U.S. and the USSR. Gagarin's orbit of the Earth propelled the ongoing competition in space between the superpowers onto a new level. The young Russian was lauded at home and abroad. His first trip outside the Soviet bloc took him to the UK, where in London and Manchester he met with the Amalgamated Union of Foundry Workers (Gagarin had been a steel worker before joining the Air Force) at their invitation. There he was greeted by thousands on the streets, and fifty years later the esteem in which he was held by the British people was resurrected when a statue to him was erected in central London. Khrushchev, already employing and indeed driving the Soviet space programme for all of its political worth, was cock-a-hoop after Gagarin's triumph. Space was now the preserve for all Soviet people. Communism had truly delivered for the masses. Less than three weeks after Gagarin's mission, however, the U.S. launched Alan Shepard on a brief sub-orbital flight to become the first American in space. He was celebrated and cheered across the U.S., showing President Kennedy, like Khrushchev, the domestic political capital that could be earned with a national astronaut programme. Kennedy's presidency had been struggling following the failure of the CIA-backed invasion of Cuba just weeks earlier, and he was now doubly alert to subjects that could improve his image (as well as his nation's) and political standing. With the cheering for Shepard still ringing in his ears, Kennedy thought once more about the political capital of space exploration. He had already asked Vice President Johnson whether the Soviets could be beaten in space and if so how. At the start of his presidency Kennedy had been lukewarm on the space programme but was now moving towards making a major commitment to it. He duly challenged the nation to deliver Project Apollo and its newly assigned mission of landing a man on the Moon and returning him safely to the Earth by the end of the decade. The Mercury manned spacecraft programme that Shepard had launched was now to be followed by Gemini, in which pairs of astronauts would orbit the Earth practising the techniques that would be required to send manned spacecraft to the Moon.

VON BRAUN'S PARADIGM

While many space advocates were enthused by Apollo, some had mixed feelings about it being catapulted to the front of the space programme. Wernher von Braun, the man who was put in charge of developing the rockets that would take astronauts to the Moon, was one such critic. He had his own paradigm of space exploration dating back to the 1950s and earlier: building a space station orbiting the

Earth from which missions could be launched to Mars and the Moon. Von Braun lived long enough to witness NASA's Skylab space station programme between 1973 and 1974, and President Nixon's decision in January 1972 to fund development of a space shuttle that could be used to provide routine and cheaper access to Earth's orbit; the Space Shuttle programme was eventually used to help construct the International Space Station. But his long-held belief in building a space

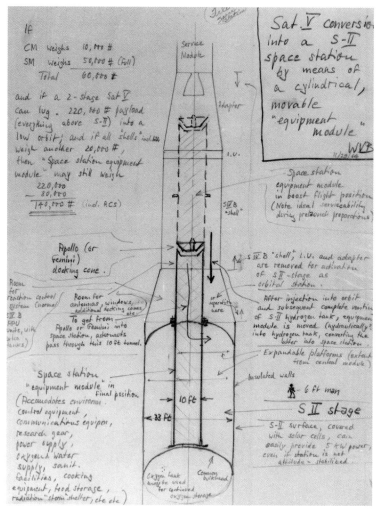

Wernher von Braun's sketch of a space station based around a Saturn V rocket second stage, 1964.

The Skylab space
station photographed
over the Amazon
from the Skylab 3
Command/Service
Module, 1973.

infrastructure of shuttle–space station–Mars (nuclear) rocket has yet
to be realized. Between 1973 and 1998 the U.S. launched no space station modules into Earth's orbit. An earlier U.S. Air Force programme to
use Gemini spacecraft as the basis of a reconnaissance Manned Orbital
Laboratory was cancelled in 1969 when the costs could not be justified
against the improved capabilities of unmanned military satellites. The

A 1960 concept image of the United States Air Force's proposed Manned Orbital Laboratory (MOL) intended to test the military usefulness of having humans in orbit. The programme was cancelled in 1969.

Nine of the thirteen astronauts and cosmonauts on board the International Space Station, 2009. Seated at the table, clockwise from bottom left, are European Space Agency astronaut Frank De Winne, astronaut Christopher Cassidy, along with astronauts Mike Barratt and Tim Kopra and Canadian Space Agency astronaut Robert Thirsk and astronaut Mark Polansky. From left to right at top are Japanese Aerospace Exploration Agency astronaut Koichi Wakata and Russian Federal Space Agency cosmonauts Roman Romanenko and Gennady Padalka.

The International
Space Station
photographed
by an STS-132 crew
member on board
the Space Shuttle
Atlantis, 2010.

Soviets, however, their own secret manned lunar programme unravel-
ling steadily after *Apollo 11*'s landing on the Moon in 1969, proceeded
with an increasingly ambitious space station programme. From 1971
they assembled a succession of space stations in Earth's orbit: Almaz,
Salyut, Mir and then the ISS (International Space Station) with the
U.S. and other international partners. The early stations' main purpose
was military, but this faded as the capability of their own unmanned
reconnaissance satellites improved. Cosmonaut crews of the Salyut
and Mir stations gradually clocked up months and years of living in the
weightless conditions of Earth's orbit. Guest cosmonauts from across
the Eastern bloc and from other friendly nations joined their Soviet
colleagues as Premier Brezhnev, like Khrushchev, used the programme
for political advantage. Later, as Soviet funds for space grew scarce, it
was decided that fee-paying space-tourists could fly up into orbit as
well. Following the fall of the Soviet Union, the Russian Federation

A full-sized model
of the Salyut 7 space
station, with a Soyuz
spacecraft docked at
the front port and a
Progress spacecraft
at the rear port,
in the Exhibition
Park of Soviet
National Economic
Achievement (VDNH),
Moscow, 1985.

worked ever more closely with the U.S. and, with the Shuttle retired in
2011, U.S. astronauts could only get to the ISS by Russian Soyuz – the
most successful manned satellite in history.

Both superpower space races had been won and lost: the Soviets
putting the first satellite into space; the U.S. landing the first human on
another celestial body. By the early years of the twenty-first century the
sparkle of political prestige was fading from the space programmes, yet
the rise of the unmanned satellites continued unabated.

The Mir space
station from the
Space Shuttle
Atlantis, 1995.

4
SATELLITES FOR ALL

In 1960, one and a half years after Project SCORE, Dwight D. Eisenhower's voice descended once again from the heavens, this time as a reflected signal transmitted from Earth and bounced off the gigantic Echo 1A satellite. This was the first of two inflated balloon satellites (team members referred to them also as 'satelloons') designed to test the use of microwaves for satellite communications. Microwaves' shorter wavelengths mean they can be more easily directed and focused, providing stronger signals over the great distances through space to future geostationary satellites. Echo 1A had a diameter of over 30 m; Echo 11 – launched four years later – was far larger at 41 m. Both were made of very thin metallized plastic that could bounce microwaves back down to Earth, acting like satellite 'mirrors'. Echo 11 could be seen easily with the naked eye, even in daylight. Arthur C. Clarke said that his early conversations with Stanley Kubrick over a project that would eventually become the film *2001: A Space Odyssey* were energized by their looking up and seeing Echo when walking the streets of New York.

The Echo 11 balloon satellite undergoing tensile stress tests prior to its 1964 launch. It was then collapsed and stowed into the small container at the feet of the man in the foreground.

Echo 1A and 11 were 'passive' communications satellites, simply redirecting the original signal transmitted from Earth. The satellite engineers' aspiration was for 'active' satellites that would intervene in the relay of signals: receiving the signal transmitted from Earth, amplifying it and then re-transmitting it instantaneously back down to its destination's receiving station (on a different frequency so that there would be no interference between the 'uplink' and 'downlink' signals).

Melba Roy, head of NASA's satellite tracking team, 1952.

Arthur C. Clarke (right) with Stanley Kubrick, standing on the set of *2001: A Space Odyssey*.

NASA's first administrator T. Keith Glennan shows then-Senator Lyndon B. Johnson, Chairman of the Senate's Aeronautical and Space Sciences Committee, a sample of the aluminized Mylar film used to fabricate the Echo 1A balloon satellite.

This mode of functioning would be equivalent in part to that used by the submarine telegraph and telephone cables, where periodically placed 'repeaters' boosted the electrical signal so that its strength would not have diminished too much by the time it reached its destination.

It was anticipated that the first communications satellites would be placed in low, sometimes called 'minimal', orbits by the newly available missile-based launch vehicles. The technologies of the time and the rockets' power were insufficient to aim for higher orbits. Launching to these lower altitudes meant the satellites would have to be travelling faster to counteract the stronger pull of gravity: the gravitational attraction between the Earth and the satellite is inversely proportional to the square of the distance between them, or in other words, if the satellite were to be orbiting at twice the altitude then the gravitational attraction between the Earth and the satellite would fall by a quarter. This meant satellites orbiting at lower altitudes would pass across the sky and over the horizon more quickly, limiting the time for transmissions.

But as more powerful rockets became available, so higher orbits could be reached where the speed necessary to maintain the satellite's altitude and orbit would be slower on account of the weaker pull of gravity. Were they to be visible from the ground, then a sequence of satellites each launched into progressively higher orbits would respectively appear to move more slowly across our skies until they just inched their way along. As mentioned earlier, by the time a satellite was launched into Clarke's 36,000 km orbit, it would not appear to move along at all; it would remain stationary in our sky as the Earth turned with it.

TOWARDS CLARKE'S ORBIT

The concept Clarke had outlined in his *Wireless World* paper in 1945 was sound, and it influenced many engineers and scientists working on the early communications satellites. Those in the satellite businesses started referring to this crucial orbit as 'The Clarke Orbit'. Interestingly, one important player who had not come across Clarke's paper was Bell Telephone Laboratories' John H. Pierce, who worked on Echo and had himself suggested the development of communications satellites in the early 1950s. Later he had pointed out the raft of technological developments, besides powerful rockets, that were easing the path to viable communications satellites. Bell had already invented the transistor in 1947, a smaller, lighter and less power-hungry switching device than the bulky vacuum tubes then being used. Bell Laboratories had also developed the first practical solar cells that could convert solar radiation directly and efficiently into electricity and would be ideal for use on satellites. Pierce worked also with colleague Rudolf Kompfner on a type of microwave amplifier called the 'Travelling Wave Tube' that could be used both on the ground and in the satellites themselves to boost the strength of signals.

Despite the appeal of geostationary communications satellites, some players such as Bell were uneasy about the technical challenges involved, and also the inevitable delay that would occur in satellite telephone conversations whose carrying signals would have to travel about 72,000 km from speaker to listener rather than the few thousand for satellites on lower orbits. Geostationary satellite communication would bring with it a good half-second between one person speaking and the other responding. Bell Laboratories focused their research on developing an active communications satellite that would orbit far lower, be less technically challenging, could be launched with existing rockets and could relay telephone calls with no delays in the conversations.

TELSTAR

With the onset of the 1960s, governments and industry either side of the Atlantic were anxious to develop satellite communications systems. They would boost industry and provide a greatly improved means of communicating over long distances. In 1962 the respective governments built large ground stations in the U.S., France and England. Each sported a massive aerial that could be rotated and tilted to follow communications satellites across the sky. The construction in England was sited at Goonhilly Downs in Cornwall. It was a massive, 26-m-diameter parabolic dish – the first satellite antenna of its kind – weighing well over 1,000 tonnes and named 'Arthur', a nod to the county's Arthurian legend. Over in the U.S., the one at the small town of Andover, Maine, was in the shape of a horn, designed to restrict the amount of extraneous radio noise interfering with the very weak satellite signals. These stations were constructed as part of an international experiment to transmit the first live television broadcasts across the Atlantic.

The satellite to be used was called Telstar, built by Bell Laboratories, and was launched by a U.S. Thor Delta rocket in July 1962. As the satellite rose 6,000 km over the Atlantic, the wheels and motors of the ground stations at Andover, Pleumeur-Bodou in France and Goonhilly Downs whirred and rumbled as they turned their antennae

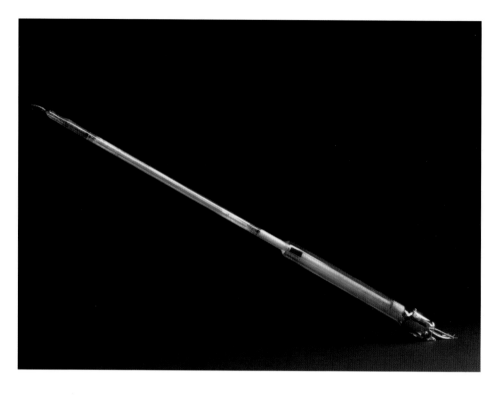

to track the satellite. The first pictures to be transmitted across the
Atlantic showed a billowing Stars and Stripes beside the massive
dome covering the Maine transmitter. They were received successfully
at Pleumeur-Bodou, but some faulty settings at Goonhilly Downs
meant the UK had to wait before it too received the first transmissions.
Nevertheless, the impact in Britain and elsewhere was profound. For
some Telstar is etched into the memory by way of the eponymous
pop instrumental that topped the charts in both the U.S. and the UK.
Even the Queen referred to the satellite in her Christmas message to
Britain and the British Commonwealth that year.

First sealed-off travelling wave tube, made by Rudolf Kompfner, England, 1945–6. The Telstar 1 satellite used a travelling wave tube to amplify the signals being relayed across the Atlantic.

Replica of the Telstar 1 satellite, 1962. The first satellite to transmit live television between the United States and Europe.

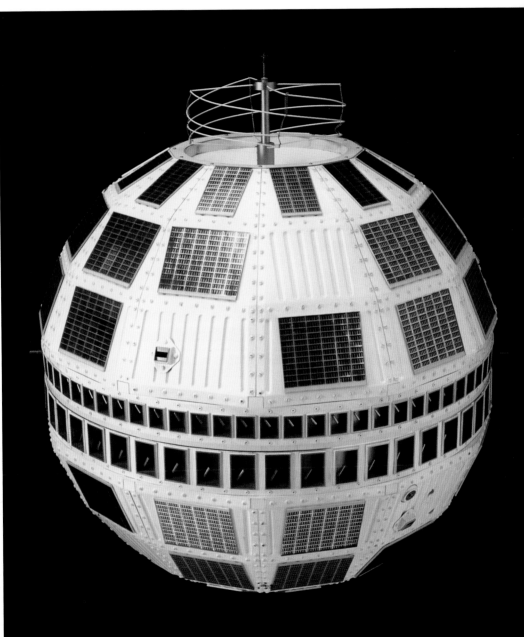

Horn antenna in Pleumeur-Bodou, France, used for the first live transmission of television from the USA to Europe, 1962.

FIRST PICTURES FROM TELSTAR

Mr Fred Kappel, Chairman of the American Telephone and Telegraph Corporation, maker of the Telstar 1 satellite, during its first live transatlantic television broadcast, 1962.

Telstar MASER
assembly with
magnet, made by
Mullard Research
Laboratories, Redhill,
Surrey, England,
1962. MASER
stands for microwave
amplification by
stimulated emission
of radiation. It was
used at the Goonhilly
satellite station in
Cornwall to amplify
signals from the
Telstar satellite.

GEOSTATIONARY SYNCOM

In 1959 Harold Rosen, an engineer at the Hughes Aircraft Co. in the
U.S., who considered his experience in designing anti-aircraft missiles
relevant to the challenge of building a geostationary satellite, and a
small group of colleagues had set about designing a satellite that could
function in the Clarke Orbit. It would be a struggle with funds in
short supply and little interest from the Hughes Company or anyone
else; U.S. rockets were still tending to explode on the launch pad.

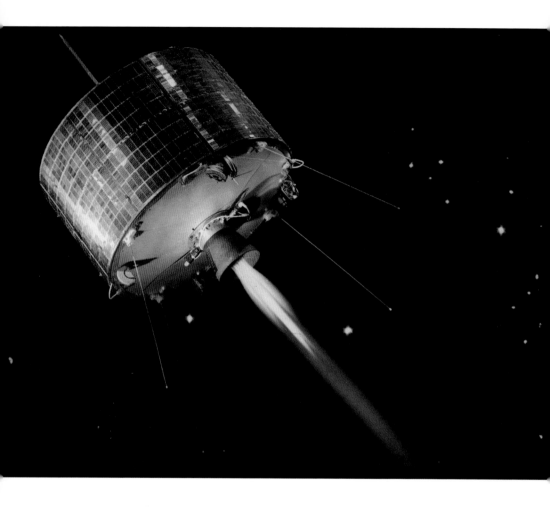

A co-worker of his built one of the satellite sensors in his garage. Eventually, Hughes and then NASA backed the geostationary concept. After one failure (Syncom 1) and one successful geosynchronous (orbiting the Earth in one day but in an orbit that is not exactly over the equator) insertion of the Syncom 2 satellite, Syncom 3 was launched in August 1964 into a geostationary orbit. It was placed so it would 'hang' over the International Date Line, mid-Pacific, a return of sorts to the Brick Moon of Edward Everett Hale (discussed in Chapter One). It had sufficient capacity to carry television signals and

Syncom 2, the first geosynchronous satellite, c. 1960. The next satellite in the series transmitted live coverage of the 1964 Olympic Games in Tokyo to stations in North America and Europe.

was used to transmit live broadcasts of the Tokyo Olympic Games to viewers across America. Clarke's vision was now being realized. Syncom was small: a simple-looking drum that measured less than 1 metre in diameter. Today's equivalent geostationary satellites can be far larger – many metres in size and several tonnes in weight. They too may be relaying television signals from one ground station to another (including those in living rooms via the satellite dishes on the sides of homes), while others, increasingly, can be transmitting huge quantities of broadband data, perhaps to users in remote areas where cable and mast infrastructures are impractical and too costly to instal.

Launching satellites to this Clarke orbit meant that the transmitters and receivers on Earth would not need to move and swivel to track them as they moved across the sky, as was the case with Telstar. The transmitter/receiver dishes could be fixed easily and cheaply to the ground or building or some other anchored structure. All that was needed then was for the dish to be pointed to the one spot in the sky where the satellite could be found. When direct broadcasting satellite

Eurostar 3000 satellite including main satellite bus and two antennae, made by Airbus Defence and Space, formerly known as EADS Astrium Ltd, Stevenage, 2000.

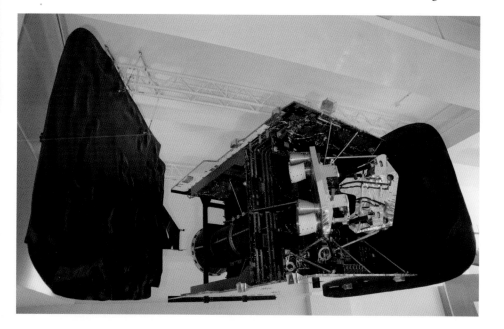

(DBS) services were being planned in the late 1970s, the prospective companies knew domestic satellite dishes would need only to be simple affairs bolted to a chimney stack or the side of an apartment building with a coaxial cable running into the room containing the television receiver. The engineer would arrive to set up the DBS equipment, locate the distant satellite 36,000 km out in space, carefully align the dish so as to point in precisely that direction and television pictures would then be received.

EARLY BIRD

The first commercial geostationary communications satellite – called Intelsat-1, otherwise known as Early Bird – was launched in 1965. This small spacecraft, no bigger than a domestic dustbin, provided the first near instantaneous commercial telecommunications service between North America and Europe. This was the mission that brought the slogan 'Live by Satellite' into the Western world's television parlance. Its Early Bird moniker came from a member of COMSAT (Communications Satellite Corporation) when he was asked by a reporter how this satellite would differ from others launched in the future. The representative predicted that as this satellite could be the first of many, it could, presumably, help corner the market (the early bird catching the worm). The name stuck. COMSAT had been created as a result of the U.S. Communications Satellite Act of 1962, which was passed to help facilitate the commercialization of satellite communications. In 1964 the Corporation helped form the International Satellite Consortium (Intelsat) and a year later launched Early Bird.

Early Bird was joined in geostationary orbit by NASA's more prosaic ATS-1 satellite in 1966. A year later these and two other Intelsats (II F-2, Lani Bird; and II F-3, Canary Bird) worked together in a very public demonstration of their function. They were the principal players of the 'Our World' television broadcast that was viewed around the world via these satellites (and terrestrial microwave links) on 25 June 1967. The

Engineers Stanley R. Peterson (left) and Ray Bowerman check the Early Bird satellite, c. 1965.

live programme aired simultaneously in 31 countries and incorporated live feeds from fourteen of them.

The programme was the brainchild of the British Broadcasting Corporation (BBC) producer Aubrey Singer, who, having worked at the UK-end of the Telstar transmissions in 1962, was well aware of the potential of the then soon-to-be-launched Intelsat satellites. They would enable a television programme to be beamed around the world, as had been predicted by Arthur C. Clarke. Public broadcasting television had existed in Britain since before the war, but by the early 1960s it had started to expand rapidly. This was partly the result of the new electronic technologies becoming available and also, in the case of the BBC, a response to the new competition from independent commercial television. The BBC had been at the forefront of television's evolution

during the 1950s, and its highly experienced staff and forward-thinking management were keen to wring the most from the new opportunities. With communications satellites soon to be added to television's armoury, here was an opportunity to create the world's first global programme – one that would be seen by millions of people around the planet at the same time. This would be the satellite as technological wonder and exemplar.

Singer and his colleagues wanted to parade this new extraterrestrial technology, albeit as an extension of television's power. The BBC's anchorman in the studio was Cliff Michelmore, and he stressed this capability during his introduction to the programme:

> [T]elevision can beat the sun. Our cameras can be where it is morning and afternoon, dawn and sunset, today and tomorrow, all with the press of a button. The dawn creeps round the equator at a mere one thousand miles per hour but our pictures flash round at 186,000 miles a second.

The satellite was the premise for the programme but was itself quite invisible. Its presence was alluded to but could not be seen. This physical absence was compensated for with allusions of planetary space in the BBC's main studio: a large globe was illuminated with a single spotlight and an elevated TV camera peered close by as if a satellite in its orbit.

One day in summer 2011, the writer Sir Antony Jay explained to me how he had suggested to Singer a theme for the programme of global population growth. The programme would employ the power of technology to sample live, real-time scenes from around the world among its constantly growing millions, the implication being that this ever-increasing rise in numbers was unsustainable. The sequences were arranged in topics: art and culture, science, sport. There was a TV crew capturing the cries of a newborn baby in a Tokyo hospital; ranchers could be seen rounding up their heads of cattle in Canada;

Intelsat IVA F-1 communications satellite prior to launch, 1975.

and in London The Beatles performed a new song they had composed especially for the programme – 'All You Need is Love'. Jay mentioned in passing how it had taken almost 100 phone calls to get The Beatles signed up to the programme. Here was the satellite as global eye, the harbinger of instantaneous mass international communication, but also as a projection of 'soft' Western power. A little over a decade later and international agreement awarded each signatory country five channels for the direct broadcasting of television by satellite. By the 1990s the first DBS companies were operating and tempting customers away from the terrestrial broadcasters. The age of the fixed domestic satellite dish had arrived. Today there are scores of direct broadcasting satellites populating the geostationary orbit high above the equator.

POLAR PROBLEMS

Clarke realized that three satellites in a geostationary orbit could not provide line-of-site television broadcasting to the entire world. At far northern latitudes and at the poles, the satellites would be very near the horizon and their signal strength proportionately weak. Soviet Russia, with its northern territories poorly served by geostationary satellites, overcame the problem by placing its communications satellites in another type of orbit, the Molniya orbit (Russian for lightning), that could deliver strong signals to the country's high latitudes. Telstar had been launched into an 'eccentric' or highly elliptical orbit which took it high over the Atlantic in a vast looping path then around to the opposite side of the Earth in a lower, faster track. In satellite terms its orbit was said to have an apogee (farthest excursion from Earth) of 6,000 km and a perigee (closest approach to Earth) of about 1,000 km. It meant that its time spent over the Atlantic was maximized (so allowing a greater period of transmission and reception of the television signals), while that time spent out of range on the other side of the world would be minimized. Satellites occupying Russia's Molniya orbits performed a similar lopsided flight path around the planet. Each satellite spends an extended time over the northern Russian land mass and a far shorter time while travelling around the opposite side of the Earth. With three such satellites each launched into three synchronized Molniya orbits, any one transmitter or receiver in northern Russia always has sight of at least one of the satellites and, with the necessary synchronization of signals, an unbroken line of communication.

During the mid-1990s satellite companies aimed for a truly global reach with a new type of communications satellite system. The Iridium consortium, for example, launched a total of 66 satellites into orbit in a little over one year between 1997 and 1998. Each satellite followed an orbit that took it over the poles at a height of 800 km with the orbits arranged symmetrically around the Earth. This allowed anyone's Iridium phone handset with a clear view of the sky to be in signal reach of at least one satellite at any one time. The satellites had inter-satellite links enabling them to relay incoming transmissions from a caller at one point on Earth via the satellite constellation and back down to the receiver at another. Iridium and similar systems were intended to provide a mass civilian global mobile phone network. But the rapid growth in the terrestrial mobile phone market soon after the Iridium system became operational proved too competitive, and the business had to be repurposed for niche users including the U.S. military, the media and explorers.

Clarke's geostationary orbit now harbours hundreds of satellites. The communications ones continue the job started with Early Bird, but with vastly increased capacities that can deal with thousands of channels of high bandwidth data. Successors to the early weather satellites are also stationed around the orbit, each trained on a hemisphere of the Earth. The military equivalents of these are here too, along with others that monitor communications, signals traffic, nuclear tests and missile launches.

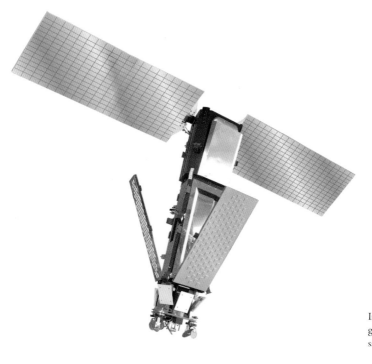

Iridium first
generation
satellite, *c.* 1997.

Iridium satellite
orbits around the
Earth.

GPS

Perhaps the most ubiquitous and widely known of satellite uses in the early twenty-first century is GPS (Global Positioning System) satellite navigation. The GPS constellation of satellites – used and developed by the U.S. military but now crucial to so many ways of life – sits some 15,000 km below satellites occupying the geostationary orbit. One of the fathers of GPS, Bradford Parkinson, told an audience at Stanford University in California, 'I don't have a prayer of telling you all the applications [of this system]', and singled out just a few in the fields of timing standardization, aviation, emergency alert, response and relief, search and rescue, agriculture, recreation, wildlife tracking, surveying and scientific research. GPS now comprises 31 satellites arranged in six planes equally positioned around Earth, each plane containing four to five satellites (the exact quantity depending on the number of reserve satellites – ready and waiting in orbit – that need to be repositioned at any one time). The familiar online animations used to represent and illustrate their locations and movement show them flying around Earth as if electrons around a nucleus, a throwback to the early signifiers of the space age. This orbital arrangement allows for any GPS receiver on Earth to access the radio signals from at least four of the satellites. GPS has become a ubiquitous component of modern life, as likely to be found in a personal handset or domestic car as in an airliner or farming tractor. The system was developed, however, as an ultimate guarantor of geopolitical power, providing U.S. submarines still more accurate coordinates from which to launch their missiles than those available with the earlier systems, including ARPA's original TRANSIT.

On Labor Day weekend in September 1973, Parkinson and a small number of fellow engineers and scientists had met at a near-deserted Pentagon building to decide on the basic form the GPS system should take. There were many options available for the system architecture, but the group decided the most challenging one to develop would offer the best solution. It would be a passive system in which an infinite number of user units on Earth could receive signals from the satellites, each user unit containing a simple crystal atomic clock with which the positioning and timing calculations could be calculated. Four satellites would need to be in view of the user on Earth at any one time and each would contain a space-hardened atomic clock. Space-hardened electronic technologies are able to withstand levels of ionizing radiation, as found naturally in space, that would otherwise damage or destroy conventional components.

The really clever bit of the system was the devising of a coding technique that enabled all four of the satellites in view to broadcast on the same frequency, yet for each still to be recognized by the receiver unit. Other challenges included designing satellites with a lifetime of at least twenty years so that fewer needed to be launched, and of assigning an affordable price tag to the receiver units that, at first, had been costing U.S.$250,000 each. The first test satellite (Block 1-type) was launched in 1978, and a full constellation was orbiting by 1985. In 1983, following the Soviet Union's shooting down of an airliner which had strayed by accident into USSR airspace, President Ronald Reagan confirmed that elements of GPS would be available for civilian use once the system was fully operational. This and President Bill Clinton's later ordering of the system's inbuilt scrambling to be turned off, so increasing the accuracy available to civilian users, allowed almost the full potential of GPS to become truly available to all.

SATELLITES AT WAR

GPS came of age in 1990 during the first Gulf War. In the 1960s and 1970s the satellite – communications, reconnaissance, meteorological, basic navigation – had been an important auxiliary tool for the military. By the beginning of the 1990s it had become an indispensable component of the military machine, no less important than the weapons themselves. The nine fully operational GPS satellites (Block II series) were launched in the course of twenty months between February

DO NOT
TOUCH

DELICATE
INSTRUMENT

1989 and October 1990. In prosecuting the Gulf War the U.S. military, according to an official assessment of its operations of January 1992, 'used space systems at an unprecedented level', and they 'supported every aspect of planning, control and execution of the war'.

Communications satellites were also of paramount importance with existing military capabilities supplemented by the repositioning in space of satellites belonging to other agencies, standby satellites brought into operation and commercial satellite capability rented out. Earth-observing satellites were vital, too. The Landsat programme has used a series of satellites from 1972 to return millions of images of the Earth's surface for use in a wide range of applications. It began scanning the Gulf area on 2 August 1990, the day of the Iraqi invasion of Kuwait, at the request of the U.S. Army, whose existing maps of the area were elderly and out of date. Weather satellites were pressed or redirected into service. The military's own DMSP (Defense Meteorological Satellite Program) satellites were repositioned from their existing orbital locations. Other weather satellites were requisitioned from civilian agencies in the U.S., Europe, Russia and Japan, or new ones were launched. The infrared sensors of the Defense Support

TRANS-PAC Global Positioning System receiver from Trimble Navigation Europe Ltd, retrieved in working order from a helicopter shot down in the Gulf War, 1991.

AN/PSN-8 Manpack
GPS receiver, one of
the first GPS receivers
used by the military.
Manufactured by
Rockwell Collins,
United States,
1988–93.

Program satellites provided early warning of missile launches. When
Scud missiles started to be launched at Tel Aviv, it was the notice
provided by the U.S. of the rockets' imminent arrival that persuaded
Israel not to retaliate.

It was to be the fledgling GPS service that proved revelatory. When
the ground offensive was launched, coalition forces were equipped
with early GPS receivers to aid navigation in the featureless terrain.

Older navigational techniques had always proved unreliable in such an environment, whether because of poor mapping, difficult weather conditions or shifting sand. GPS proved invaluable from the outset, leading rapidly to a shortage of receivers. The military was forced to appropriate the civilian sets recently placed on the market for use on yachts and motor boats. Equivalent units were pressed into service for the air force with improvised receivers lashed into cockpits. Experimental GPS units were fitted to Tomahawk cruise missiles.

GPS became an intrinsic part of modern 'smart' weaponry. The first Gulf War was, indeed, a landmark in the widespread use of different satellite systems as vital components, not only for the military but for the press and media. Correspondents embedded with the ground forces beamed live reports on the hostilities to millions of living rooms around the world. The military held press briefings to explain to global audiences how precision weapons were being used on the battlefield for the first time. For one corporation – CNN (Cable News Network)

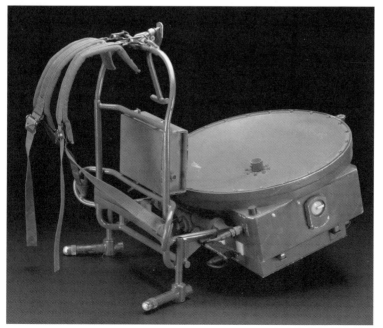

Prototype Manpack ground terminal developed for experimental use by Royal Signal and Radar Establishment on the UK MOD Skynet satellite communication system, manufactured by Ferranti Limited, in Poynton, Cheshire, England, 1980–89.

Magellan 1000M
GPS military (left)
and 1000 GPS civilian
receivers, made by
Magellan Systems
Corporation,
California, United
States, 1990.

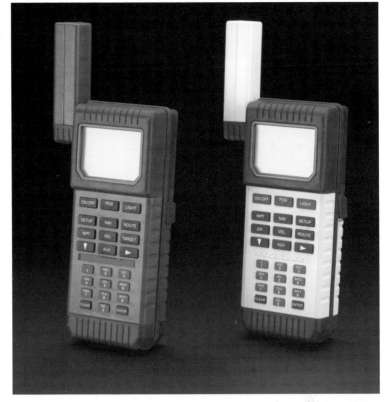

– the war became a defining moment; its reporters were the only ones from any of the major U.S. networks to broadcast, via satellite, live from Baghdad during the coalition forces' opening bombing raids of the war. Their live bulletins became a staple television diet, the first time warfare had been witnessed remotely, in real time, by civilian audiences. A visit to the BBC TV Centre shortly before it closed in 2013 revealed a room full of redundant satcom equipment used by its own reporters in the field during and after the conflict. The satellite had become a vehicle for conveying power, both hard (from the military) and soft (from the media with their reporters in the field of operations).

The war was an overture to the extensive use of increasingly sophisticated satellite services across society. Virtually all of society's critical infrastructures – water, fuel and electricity supplies, health, transport and communications networks, government, security and military systems, banking and economic facilities – came to depend to a greater or lesser extent on an orbiting space-based component operating seamlessly alongside the terrestrial infrastructure. For example, energy companies' monitoring of existing oil and gas pipelines and the construction of new ones came to depend on a cluster of satellite provisions: Earth observation and GPS for mapping and locating, and communications for the teams in the field to maintain contact with each other and their headquarters. Indeed, exploration concerns depended on satellite surveys of the Earth's surface to gather evidence of new reserves, while meteorological satellites would be used to provide weather and climate data for the oil distributors to predict temperature changes more accurately and so plan the supply of oil more efficiently.

The satellite as component remained in the hands of the traditional elites – the government agencies, the military, the scientists and the

Full-size model of GPS Block IIF satellite model, made by Paragon Creative, York, 2014.

 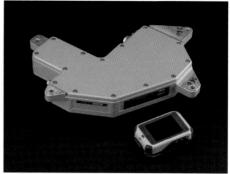

Micro Defence Advanced GPS Receiver (MicroDAGR), with colour touch screen, built-in maps and anti-jamming technology. These receivers started being used by NATO forces from 2012. Made by Rockwell Collins, 2012.

Collins TLAM GPS receiver (right) as fitted to Tomahawk cruise missiles. Rockwell Collins, 1989.

commercial operators – but was now becoming available to the ordinary user as well. While the car sat-nav became de rigueur, high-resolution satellite images of locations and roads revealed for anyone not just where things were but what they looked like. And with views from space stitched into images captured on the ground users were able to zoom down from space and onto the street.

5
SCIENCE OF SATELLITES

How does a satellite work? How is it that once launched into orbit by a rocket it can continue to freewheel around the Earth for days, months, years, even millennia? How does one actually go about launching a satellite? And what happens to a satellite when it comes to the end of its life?

In 1990 I put together a small exhibit at London's Science Museum called 'Spaceplanes'. It featured new (and some old) concepts for craft that could – in theory – take off from the ground, much as an aeroplane does, but then ascend and accelerate still further to reach space. The unifying concept of many designs was that they would each be single vehicles, not split into stages like today's conventional rockets. Each would return intact from space. At the moment the technology to build these is not there. No single, entire rocket vehicle is capable of achieving orbital velocity. It would be too heavy. The huge quantities of propellant that are needed to accelerate it to 8 km per second would require vast tanks that would in turn necessitate even bulkier support structures, and this bulk would require yet more propellant, which would require even larger tanks . . . and so on.

In the early twentieth century Konstantin Tsiolkovsky had been well aware of this fundamental problem and had written about a staging system in which each space rocket would actually comprise several independent rockets or stages. The first stage would fire and lift the entire rocket off the pad and accelerate it to a specified velocity and altitude. When that stage had exhausted its propellants, it would be jettisoned and a second stage ignited. With the overall vehicle now far lighter, the second stage would be able to accelerate the rocket further. Additional stages would be used as required to continue the acceleration until orbital velocity was achieved. Tsiolkovsky called these staged rockets his space rocket trains. Today the staging system is an essential feature of all space rocket design, with each satellite launch depending on the sequential firing of a rocket's constituent stages. Scientists and engineers have long thought about and worked on new propulsion and materials technologies that would enable a single-stage-to-orbit (SSTO) launch vehicle to be developed – one that would launch (and return) as a single unit. To date the costs of overcoming the associated technical challenges dwarf even the high costs of using conventional rocket launch vehicles, and the 'expendable' way of reaching Earth's orbit and beyond remains the only way.

The spent stages of today's rockets can join the satellite in orbit or fall back to the ground or ocean. Many launch vehicles augment their liquid-propelled engine stages with solid rocket boosters, such as those that were used by the Space Shuttle. These are enormously powerful and provide most of the thrust at launch and during the early stages of the ascent. They are essentially cylinders filled with rubbery, mouldable propellant – usually a mixture of aluminium powder and ammonium perchlorate held together with a binding agent. Following retrieval these boosters can usually be refilled and used for further launches.

The rocket's stages are held together by explosive or pyrotechnic bolts programmed to detonate and break at certain moments of the ascent, allowing the spent stage to be cast off. The jettisoning can be actively assisted (to ensure no risk of collision between the stage just spent and that about to ignite) by way of small rocket thrusters located on the spent stage that point back and brake its motion. In 1971 the UK's Prospero satellite had one of its four aerials damaged shortly after entering orbit when the final stage of its Black Arrow launching rocket collided with it.

The Space Shuttle's
solid rocket booster
falling away during
the first launch of the
Columbia vehicle,
1991.

Almost all launch events during the rocket's ascent are controlled
by on-board computers. These control systems remain similar in principle, if far more sophisticated, to those first demonstrated on the v-2
rocket, which used analogue apparatus to adjust the missile's flight.
Each v-2 was preprogrammed to hit its chosen target. Once airborne
the rocket's gyroscopes, spun up before launch, would remain spinning in the same plane so that any unexpected change of direction
of the rocket (caused, for example, by wind gust), could be measured

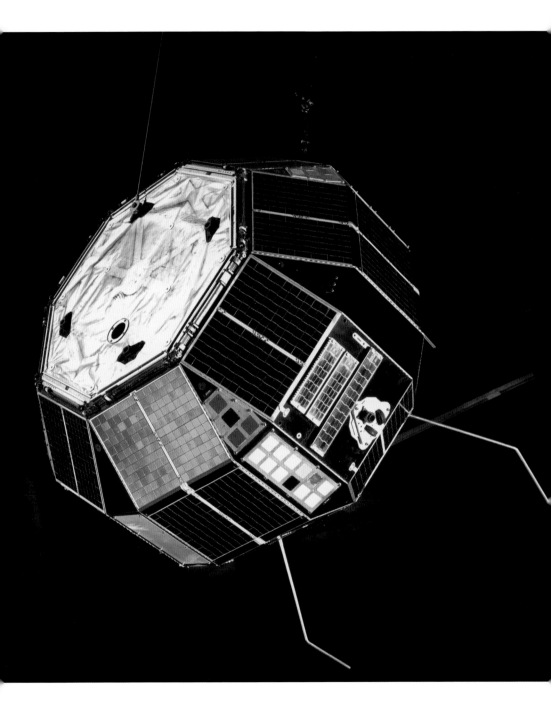

The Prospero (x3) satellite flight spare in the Science Museum's Exploring Space gallery, 2000.

Scale model of a proposed design for the UK's x3 satellite by Hawker Siddeley Dynamics Ltd, Stevenage, c. 1970.

v-2 rocket's
inertial platform,
c. 1945.

electrically against the fixed 'frame' of the spinning gyroscopes. The
deviation from the intended trajectory would trigger electrical impulses
to the appropriate control surfaces of the rocket. A trimmer was located
on the edge of each of the rocket's four fins and worked rather like the
aileron – the hinged panel on the trailing edge of each aircraft wing
used to roll the aircraft and so alter its flight path. Beneath the rocket's
engine nozzle were two pairs of vanes so positioned that the stream of
exhaust could be deflected by them; each was made of solid graphite
to withstand the high temperatures of the gases. All of these control
surfaces were actuated by servomotors that responded to the impulses
from the gyroscopic guidance unit. Those beneath the engine nozzle
would steer the rocket at high altitude where the atmosphere is thin
and aerodynamic control more difficult.

The v-2 climbed vertically before tipping gradually to continue its
ascent in a parabolic curve. On-board accelerometers sensed when the
requisite velocity had been reached for 'Brennschluss' – the moment
for the engine to shut down and for the missile to continue its course
on a ballistic coast. Today's space rockets use far more advanced, digital

Missile-derived launch vehicles at the Kennedy Space Center, Florida. Left to right: Mercury-Atlas, Atlas-Agena, Mercury-Redstone, Thor/Delta, Juno II, Jupiter-C/Juno I with an Explorer satellite attached.

computers, gyros, accelerometers and other sensors to control the vehicles as they ascend towards their target – orbit – but remain the same in essence as those first developed for the v-2 long-range missile.

Almost all of the early satellite-launching rockets derived from adapted ballistic missiles of the 1950s and 1960s. The first satellite was launched by a converted Soviet R-7, the world's first intercontinental ballistic missile. This rocket design continued to be used in uprated forms into the twenty-first century. The U.S. modified its Redstone, Jupiter, Thor, Atlas and Titan missiles for launching spacecraft. The Atlas and Thor missile-derived launch vehicle continued as the heavy-lift Atlas V and Delta IV rockets. There were many more space launch vehicles to choose from for your particular mission. A satellite's designated orbit (which depends on the function the satellite is required

One of four solid rocket motors being lowered into position alongside its Atlas V launch vehicle, Cape Canaveral Air Force Station, Florida, 2015.

to carry out) and its weight are the two key factors affecting the choice of rocket for the satellite customer. Many became available in different versions depending on the satellite mission's precise needs. For example, the American Atlas V vehicle had a first stage using a (Russian) single, twin-chambered RD-180 engine. To this core first stage could be added a maximum of five solid rocket boosters to add to its thrust – the exact number depending on the mission's particular requirements. The Atlas's upper stage also came in options: it could be made up of either one or two RL-10A-4-2 engines. A rough rule of thumb says that the heavier the satellite and/or the higher the orbit required for the mission, the greater the total amount of thrust needed and so the greater the number of engines/boosters needed. An Atlas V effectively trebled the performance of the core first stage by adding alongside it a pair of similar liquid-propelled booster rockets.

A rocket's explosive launch and ascent disguises the intricate manoeuvres entailed throughout its climb towards orbit. During the first few moments after launch the rocket rises vertically so it can reach the thinner atmosphere as quickly as possible. Aerodynamic resistance drops with altitude and aids the rocket's acceleration. It will then start to tilt towards the direction of the chosen orbit but carry on climbing until it reaches the correct altitude. The ascent is rapid and a typical launch would have the rocket and its satellite payload achieving orbital velocity just eight minutes after launch. The final injection into orbit must be done as accurately as possible to ensure precisely the correct orbit is achieved. Once this is done the final adjustments to the satellite's positioning and orientation can then be carried out by small rocket thrusters set into the satellite itself. If the satellite is destined for a higher, geostationary orbit, a rocket engine (called the apogee motor – the terms 'engine' and 'motor' are often interchangeable) attached to the satellite can be used to lift its altitude still further. This manoeuvre can take far longer, sometimes days.

ORBITS

The Earth spins beneath as the satellite performs the same circuit in space. The satellite's plane of movement is tilted to a greater or lesser extent in relation to the Earth's equator. This is called its 'angle of inclination' and is measured in degrees. So if, for example, the satellite is moving around the Earth in a plane parallel to the equator, its angle of inclination will be 0° (and it would be in an equatorial orbit). If, however, the satellite is orbiting in a plane that takes it over the poles (in a polar orbit), its angle of inclination will be 90°. The angle of inclination is measured as the satellite crosses the equator going north. A satellite launched into an orbital plane will have an inclination between 0° and 180°; inclinations above 90° imply a retrograde (westward) orbit. Most satellites are launched in an easterly direction so that they follow the rotation of the Earth. This reduces launch costs as the speed of rotation of the Earth, which is at its maximum at the equator, can be subtracted from the total velocity the rocket needs to achieve orbit; the Earth, because of its spin, can provide the rocket with its first 1,600 km per hour (0.4 km/sec) of velocity necessary to achieve orbit, assuming the satellite is being launched from the equator in an easterly direction.

The Comet Nucleus Tour (CONTOUR) spacecraft being lowered towards its apogee kick motor in the assembly hall. Contact was lost with the spacecraft one month after launch when the motor was ignited to boost it out of Earth's orbit, 2002.

The choice of orbit is determined by the function the satellite needs to perform. The high geostationary orbit over the equator will be the destination for those satellites needing to face the same part of the Earth all the time: communications and television satellites servicing specific areas of population; meteorological satellites monitoring the weather systems of particular regions – Pacific, Atlantic, Eurasian and so on – and military satellites focused on adversaries' defence forces. Satellites using polar orbits will pass over a greater part of the entire globe, depending on their precise inclination, with the Earth spinning beneath them; after a number of orbits the satellites will return over the same points. Reconnaissance satellites also use this type of orbit, usually synchronized with the day so that the shadows cast by their targets are of similar length, making target movement and change all the easier to discern. Families of satellites launched into medium-altitude orbits of various inclinations, in which at least one will be in view of any point on Earth, are used for global, mobile communications networks.

For easterly launches it is important that the areas underneath and near to the launch and ascent of the rocket are remote with little or no population centres. The first stages of rockets fall back to Earth and are therefore a hazard for anyone living nearby. The U.S.'s Cape Canaveral on the Florida coast and the European Space Agency's Kourou facility in French Guiana both launch over the Atlantic, where intended (and any unintended) debris will fall safely into the ocean. Russia's site at Baikonur in Kazakhstan launches over largely uninhabited land.

However, the position of the site also affects the choice of orbital plane the satellite can be injected into. The laws of Johannes Kepler, seventeenth-century mathematician and astronomer, describe the motions and orbits of the planets. By applying his first law to an artificial satellite, we know that the centre of the Earth has to be in the centre of the satellite's orbital plane. As the launch site must lie on or very near to the orbital path (the rocket needs to align itself to the orbital trajectory as soon as possible after launch), then all the orbits that can be reached directly from that site must have a minimum inclination equal to

Engraving from Johannes Kepler, *Astronomia nova* (New Astronomy, 1609), in which he compares his own theories of planetary motion with those of Copernicus, Brahe and Ptolemy.

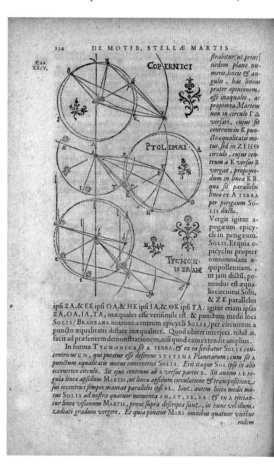

Redstone missile
CC-1002 on the
launch pad at Cape
Canaveral, Florida,
16 May 1958.

the site's latitude. So those satellites launched directly into orbit from Cape Canaveral at latitude 28.5° North have a minimum inclination of at least 28.5°.

SATELLITE STRUCTURE

Although it had long been predicted that satellites could perform a variety of jobs – meteorological, communicatory, scientific – all that Sputnik really had to do was to make it to orbit and survive long enough to be able to let the world know that it had arrived there. Such had been the intention for the U.S. Vanguard project as well, but that had only made it into orbit at the third attempt. Both the American and Soviet satellite designs were minimalistic: 'simple' casings that would primarily demonstrate an ability to withstand the violence of launch and protect the internal instrumentation from the extreme environment of space. Sputnik and Explorer were both spherical to impart maximum structural strength for the least weight. Explorer, the first U.S. satellite, was a more rudimentary design, formed from a modified Baby Sergeant rocket motor: half was filled with solid propellant and provided the boost into orbit; the other half contained a radio transmitter, scientific instruments and batteries.

Excessive weight is a fundamental challenge to all satellite design; its weight is limited to the maximum the launching rocket is able to lift into space. Lightweight metal alloys, honeycomb panels, carbon-fibre composites and plastics all came to be used to provide the greatest strength with the least weight.

HAZARDS IN SPACE

Once in space there are plenty of hazards and extremes the satellite has to be able to withstand. The visible sunlight that bathes and supports Earth is but a small part of the radiation that the Sun emits. It also gives out invisible light from across the electromagnetic spectrum – X-rays, ultraviolet, infrared and radio waves. The solar wind of mostly electrons and protons blows onto the planet, too. The wind is invisible but its effect is revealed when the particles collide with atoms of the atmosphere. These collisions generate visible light that appears to bend and bow and shimmer under the influence of Earth's magnetic field. These are the beautiful and ethereal aurorae: the borealis of the north and the australis of the south.

Much of this solar radiation is extremely harmful both to life and to the delicate electronic components used in satellites. The Earth's magnetic field and atmosphere protect living organisms from the worst of it, but for an artificial satellite orbiting high above the atmosphere, there is no such natural protection. Indeed, the Earth's magnetic field lines also trap two belts of Van Allen radiation, a fact that limited the orbiting of early satellites. If missions require the satellite to pass through the belts, then the satellite has to be suitably strengthened to withstand the radiation, in other words, its structures and components must be space-hardened to withstand the damaging radiation. Sometimes its sensitive systems and detectors will be turned off during passage. The Apollo mission astronauts passed through these belts on their way to the Moon, but their exposure to the radiation was brief and the dosage received was far lower than that experienced in many environments and occupations on Earth. Nevertheless, the space radiation could play havoc with satellite systems and instrumentation, eroding external surfaces and inducing electrical charges within circuits. A satellite must therefore be protected and shielded from such an attack. Much of the technology and techniques used to effect this hardening was originally developed to protect electronic materials and components from the consequences of exposure to a nuclear blast.

The solar wind can also affect the satellite's electronics. A satellite's orientation and motion are influenced by the momentum imparted by the wind's constituent photons. Photons may not have mass but they do have momentum; the momentum exchange that occurs when they strike the satellite's surfaces is sufficient to deflect the velocity and position of the satellite, much as the wind blows at the sails of a ship. The effect is small but accumulates with time and can be ample enough to wreck the precision of a mission. A satellite has therefore to be able to compensate for this effect by returning itself to its designated position. During periods of heightened solar activity, the satellite operators have to be particularly alert to such events, and satellites may even be shut down temporarily to protect the electronics.

Satellites will be subjected to extremes of temperature depending on whether they are in sunlight or shade, so they need to be able to withstand temperature fluctuations of several hundred degrees Celsius. Sputnik's shine helped reflect some of the solar radiation, and a small fan inside was automatically triggered if the temperature rose above 30 degrees. Later satellites started to employ passive and active environmental control systems. The passive system modifies the satellite's external surfaces by cladding them with highly reflective (gold- or silver-coloured) multiple-layer insulation sheets. These protect the satellite from excessive cooling when in shade and excessive heating when in sunlight. The active systems can include electrical heaters and heat pipes filled with ammonia or ethane; for an area of the satellite that risks overheating the pipe fluid can evaporate down the tube and condense in a cooler region of the satellite, carrying the excess heat with it.

POWER

All satellites, and indeed spacecraft, depend on a supply of electricity to power their radio and instruments. The first Sputnik carried three silver-zinc batteries, which comprised a significant proportion of the satellite's mass. They functioned successfully for three weeks before running down. Explorer carried mercury-zinc batteries, and these represented 40 per cent of the satellite's total mass. The much larger Sputnik 3 – a 3.5-m-long cone containing a dozen scientific instruments – required a whole stack of batteries: most of its interior was filled with them. A small patch of experimental solar or photo-voltaic cells was located on the panelling, and this powered one of the satellite's transmitters. Vanguard had also carried some early solar cells.

Today's satellites carry an equivalent but much more capable combination of rechargeable batteries and solar cells that ensure electrical power is maintained even when the satellite passes into the Earth's shadow. As larger and more power-hungry satellites were launched, so larger numbers of solar cells were required. Some satellites housed them on small paddles – often four – arranged symmetrically around the satellite's main body. When still greater numbers were required they were arranged on the satellite's outer surfaces, covering almost the entire satellite. Various polyhedral satellite designs evolved to increase the available surface area, each face comprising a patchwork of solar cells. The 'drum' became a widely adopted satellite shape, with the cells forming its curved surface. This shape could also be relatively easily stabilized in space. As satellite missions and functions became increasingly sophisticated, so the precise position and orientation within the orbit became critical: they needed to be stabilized so as to avoid drift and tumbling, as these types of movement would interfere with communications and disrupt the functioning of any scientific instrumentation. Stabilization would be achieved by 'spinning up' the satellite during its release from the rocket. Small rocket thrusters aligned with the circumference of spin would usually be used. The spin rate would likely be between 30 and 100 revolutions per minute (rpm) and would ensure the satellite's vertical axis was steadied perpendicular to the plane of its orbit, rather as a gyroscope maintains its orientation when spinning. Part of the satellite might need to be stationary in relation to the spinning part; it would be 'de-spun' so as to provide, for example, a fixed orientation for the satellite's antenna (which would need to be pointing to Earth). But as satellites continued to evolve in function and complexity, requiring still more power, it became necessary for far greater areas of solar cells – more than could be accommodated on just the surface of the satellite body alone.

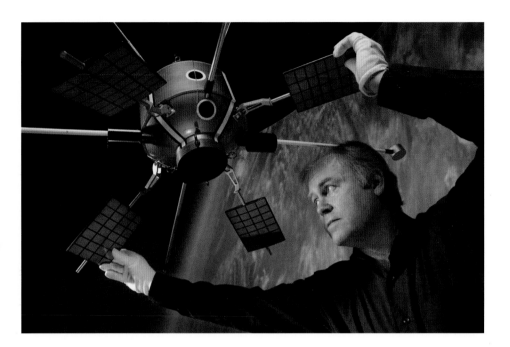

Scale model of the Ariel 1 satellite in the Science Museum's Exploring Space gallery, 2014.

It is useful to liken a satellite's sensory needs to those of a human. The human inner ear provides a sense of balance and orientation – that is, it indicates which way is up. A spinning satellite achieves this but is limited to maintaining just the one 'vertical' position, like the spinning top. In the case of a three-axis stabilized satellite, its movement can be controlled in each of three axes of pitch, roll and yaw by reference to a fixed point in space. The point may be a star, the Earth's horizon or even the Sun. Light-sensitive detectors set into the satellite respond to the source and provide a fix for the satellite to orientate itself. This is normally carried out with momentum and reaction wheels positioned within the satellite that, when set spinning, induce a rotation of the satellite in the opposite direction.

The Ariel 1 satellite flight spare in the Science Museum's Exploration of Space gallery, *c.* 1990.

This refined stabilization allowed arrays of solar cells to be used. These panels, usually as a pair, would be deployed once the satellite had been placed in a stable orbit. During launch they would have been stowed in a folded position, against opposite sides of the satellite body.

Once opened they would be extended out either side of the satellite to give it a distinctive 'winged' look. Such an arrangement of solar arrays provides an optimum configuration for generating electricity from sunlight – greater power requirements could be met simply by adding extra panels to the arrays or making each one larger. But they needed to be able to point to the Sun, and this was made possible through three-axis stabilization.

Three of the International Space Station's sixteen solar panels are clearly visible, with the Earth's horizon in the background, 2008.

Satellites may also experience slight temporal drift within their assigned orbit. This may be caused by gravitational perturbations from the Sun, the Moon or the Earth itself owing to the uneven and shifting distribution of their gravitational force. Heightened solar activity can also induce such drift. It can be corrected through the firing of small thrusters to return the satellite to its designated position. The thrusters can be simple gas jets, small chemical rockets or even electrical engines that impart movement through the ejection of charged particles.

Consigning the solar cells to remote arrays also freed up space on the satellites' faces for complex mounting structures, including large dish antennae. The early satellites received and transmitted relatively weak radio signals at low frequencies and at low gain rates; in other words, the rate of data transfer was low. Their antennae were omnidirectional and could be simple aerials. As satellites started to use a broader and higher range of frequencies for greater rates of data transfer, so the antennae needed to become larger and more unidirectional to cope with the narrower signal beams. Modern communications satellites started to use multiple frequency beams, each bouncing off a different spot on the antenna reflector to supply a wide range of signals to dispersed population areas. Reflectors were also contoured so that beams could be differentiated to focus preferentially on populated areas. Some reflectors became as large as, or far larger than, the satellite itself, as increasingly demanding data transfers on the ground were required in, for example, smaller satellite phones. Such reflectors depended on the development of sophisticated, lightweight structures that could be remotely unfolded in space.

PAYLOADS

A satellite's payload, or cargo, depends on its mission. A communications satellite's payload is a transponder or set of transponders. These relay the signals from one point on Earth to another. Each comprises an input where a signal is received in the 'uplink', amplified and

converted to another frequency for retransmission as the downlink. The frequency change ensures there is no interference between the uplink and downlink. For a high amplification the incoming signal is gathered by the reflector dish and focused onto a receiver called a 'feed horn', which conveys the signal along a tubular waveguide and into the transponder. The same components can be used in reverse for the transponder's output back down to Earth. Over time satellites started to use many transponders, each able to deal with a range of frequencies simultaneously.

NASA's NEXIS ion propulsion engine, which uses electrical power, c. 2003.

The payloads of Earth-observing satellites are the detectors sensitive to different types of radiation emitted by or reflected from Earth's surfaces or atmosphere. Some respond to visible light, and many others to infrared – all objects above absolute zero (the lowest possible temperature) emit a certain amount of infrared. One type of U.S. satellite carried a particularly large and sensitive detector that could register the heat from missile exhausts. These DSP (Defense Support Program) satellites have long been part of the U.S. early warning system against missile attack and in recent years have been supplemented with more satellites placed in highly elliptical orbits. The sensors employed are sensitive to infrared light in much the same way that night-vision cameras can detect and envisage the forms of, say, animals and people from the heat emitted by their bodies. The satellite's sensors are far more sensitive and tuned also to specific heat 'signatures', such as those characteristic of a launched missile's exhaust plume.

The satellites used to generate pictures of clouds for the television weather forecasts also carry infrared sensors as payload, this time to measure the temperature of clouds rather than that of missile exhausts. The tops of clouds are cooler than their bases; the higher the cloud, the colder its top. So the tops of higher, rain-bearing clouds will be emitting less heat than smaller ones and will show up as being cooler by the satellite's infrared instrument. The meteorologists then align these temperature data with a grey scale that shows the higher (rain-bearing) clouds as white and the lower, smaller (non-rain-bearing) clouds as grey or even translucent, and it is these images that are shown during the television weather forecasts. These satellites usually carry detectors that operate in the visible part of the electromagnetic spectrum as well, so forecasts will often combine the infrared with the visible to provide a more useful visual interpretation of the weather system in question.

Another Earth-observation technique uses radar to generate relief images of the Earth's surfaces. Unlike optical and infrared sensors that detect reflected or emitted light (including heat sources such as forest fires or artificial lighting) from the Earth below, these payload instruments generate their own illumination in the microwave part of the electromagnetic spectrum, which is bounced off the Earth's surfaces and then reflected back at the satellite. This means they can function around the clock with darkness making no difference to their efficiency. Radar can also penetrate clouds (unlike optical sensors) and is therefore unhampered by weather. Satellites carrying radar instruments are launched into lower orbits, as the reflected beams diminish quickly and would carry little useful information by the time they returned to a higher orbiting satellite. Early Soviet radar satellites could only be launched into very low orbits on account of their bulk; the rockets available were not powerful enough to loft them higher. This meant they could not employ the solar panels that are normally used to generate electricity for powering the satellites and their instruments. Microwave radar generation is power-intensive, and had the Soviets used solar arrays they would have had to be very extensive in size to generate sufficient electricity to then produce the microwaves. But their large surface area would have slowed the satellite's speed as it collided with more and more gas molecules in the upper atmosphere. This would have caused it to descend as its orbit decayed and for the solar arrays to be torn off as it dropped further into the atmosphere. These Soviet radar satellites therefore generated the necessary electrical power with small nuclear reactors.

Radar observation satellites provide good imaging of terrain and structures and so are increasingly used for military mapping. They return useful information, too, on the nature of ocean surfaces, including wave heights and currents, and have therefore been included on meteorological and climatological satellites. These types of satellite mission, therefore, tend to fly in low- and medium-altitude orbits, the exact choice of orbit a compromise that best suits the requirements of all the instruments carried.

GPS IN DETAIL

The GPS satellites' fundamental payload is an atomic clock, one on each satellite. All are synchronized, the signal from each transmitted for reception on or near Earth (sometimes by other satellites). If a receiver can detect signals from four satellites, they will each arrive at slightly different moments. The receiver's processor is able to compute these differences and so calculate its own position. It does this through a process called 'trilateration', a technique that exploits the geometry of circles and spheres. If the GPS receiver calculates the distance to one of the GPS satellites, then the receiver could be anywhere on the surface of an imaginary sphere drawn around the satellite. The receiver (and user) could be anywhere on that surface, whether on land, in the air or even in space. But if the receiver then calculates its distance to a second GPS satellite, then it could be anywhere on a circle formed in space where the two imaginary spheres intersect. The sphere of a third satellite will cross the circle at just two points, one of which will be that occupied by the receiver, so giving its position. Reception from a fourth or more satellites allows the accuracy and timing of the locating to be increased still further.

END OF DAYS

There is displayed in the Science Museum in London what appears to be a rusty old food can. It is in fact an artefact from the Lower Body Negative Pressure Experiment that was flown on the Skylab space station in the early 1970s. Astronauts 'wore', or lay within, this device to investigate the effects of weightlessness on their cardiovascular systems. It remained on board when the final crew of astronauts departed in 1974, and five years later, as the vacated space station plunged through Earth's atmosphere, broke into pieces, one of which fell to Earth in Australia and is now displayed in the gallery. All artificial satellites eventually fall back to Earth, given enough time, but many stay in orbit for what will be hundreds or thousands of years or even longer. There is much else up there, too. Earth's space is strewn with space-age debris, mostly invisible to our eyes until re-entering the atmosphere, whereupon it appears as artificial shooting stars or the occasional fireball (signalling the incineration of a far larger piece of space junk).

The accumulated space debris contains pieces of satellite and rocket that may have broken off during the ascent to orbit. There will also be the odd tool and other items accidentally dropped by astronauts and cosmonauts during their spacewalks. Most of the debris is far smaller though, but can still cause serious damage when striking a satellite or spacecraft. In 1983 a fleck of paint from a rocket slammed into the orbiting Space Shuttle's flight-deck window, leaving a crater, much as a flying stone would scar the windscreen of a car. When a similar incident occurred a few years later, NASA had to consider the risk of such an event actually puncturing the glass and causing a catastrophic decompression of the Shuttle, and thus made the decision to alter the orientation of the Shuttle as it orbited Earth. For all future missions the Shuttle would 'fly' in orbit facing backwards, its stronger underside leading and thus protecting the more vulnerable windows from potential damage. Vulnerable instruments on satellites can be afforded some degree of protection with so-called 'Whipple shields' that are positioned close to the surface to be protected and disrupt any incoming high-velocity particle. Despite the large quantity of space debris orbiting Earth, its distribution is vast, and significant impacts with satellites correspondingly rare. And individual pieces are falling back towards Earth each day, most burning up harmlessly in the atmosphere. The frequency of return increases with heightened solar activity: a more active Sun heats the Earth's atmosphere, causing it to expand, and it therefore envelops and brings down more junk.

On rare occasions, a launching rocket may explode near to reaching orbit, or there is a catastrophic collision between satellites, resulting in the generation of large amounts of rubbish. In 2009 a retired Russian satellite and an American communications satellite rammed into each other, creating a huge cloud of debris. Rigorous mission planning and international agreements normally prevent such accidents occurring; each satellite is meant to be launched such that it never occupies the same area of space at the same time as another. Nevertheless, the material from the 2009 incident will remain as a hazard in space for many years to come, and this has added to a mounting concern over the amount of junk in orbit, especially in the case of errant or dead whole satellites. In April 2013 a report concluded that there was an urgent need to start removing debris from orbit in order to avoid a calamitous cascade of collision detritus that could significantly degrade the many services that society now depends upon from satellites – communication, navigation, Earth-observation and so on.

Debris removal is difficult, with most proposed solutions involving sophisticated and challenging techniques for slowing whole dead satellites so as to initiate their subsequent de-orbit. Satellites orbiting at low altitude and nearing the end of their useful lives can be programmed to fire their small manoeuvring thrusters against their direction of flight, so acting as a braking mechanism to slow them and allow capture by Earth's gravity, pulling them down to burn up in the atmosphere. Increasingly, satellite operators are being encouraged by national space agencies to ensure sufficient propellant is retained in the satellite's tanks to ensure manoeuvrability and braking is possible before they drift. The higher the debris is orbiting the longer it will be before it falls back down into the atmosphere and incinerates. Those satellites orbiting at high altitude – especially those in the 36,000-km geostationary orbit over the equator – are now required to be boosted at the end of their lives into a slightly higher 'graveyard' orbit, where they can be safely left for many hundreds of years.

With ocean comprising four-fifths of the Earth's surface, the risk of any returning hardware hitting a populated area is low, but the debris trajectory still needs to be calculated to ensure that the danger remains minimal. The u.s. Skylab space station was coming to the end of its life in 1979, and NASA controllers were able to adjust its orientation so as to help steer its re-entry into the sea to the south of Africa. At the time there was a great media hoopla as to where, exactly, it would come down; this was by far the bulkiest 'satellite' to have de-orbited to date. As it turned out the calculations were not perfect and Skylab broke up over southwest Australia, with parts crashing to Earth near the town of Esperance. No harm was done, but the local authorities issued a fine to NASA of a few hundred dollars for littering the countryside.

Strictly speaking, any returning piece of space hardware belongs to the organization that owned it when it was flying in space. This is required of all signatories to the United Nations' Outer Space Treaty of 1967. By tracking the satellites as their orbits decay and they near re-entry, their identity and owner can be confirmed.

Active satellite removal of total dead satellites remains extremely difficult, with no proven system yet developed. The best that can be done is to track the larger pieces as they orbit the Earth so as to get warning of any potential collisions. This can be done for items greater in size than about 10 cm (roughly the size of a large orange), and the necessary planning and/or evasive action programmed into or uploaded to vulnerable satellites. Maintaining an accurate and up-to-date log of orbiting objects is a complex and costly process and is carried out largely by the military. This is done not only to learn of likely collision paths and issue alerts to those affected, but to distinguish orbiting objects from incoming enemy missiles. The u.s. Department of Defense (DOD) draws on a variety of techniques to ensure such 'Space Situational Awareness'. Satellite positional data are returned to a string of high-frequency radar installations that stretch across the u.s. at latitude 33° North between California and Georgia. There are three transmitting stations and six receiving ones. This array is capable of detecting any object 10 cm or larger and orbiting above the u.s. out to a distance of 30,000 km. A far more powerful radar system goes by the name of the wonderfully acronymic PAVE PAWS EWR (Phased Array Warning System Early Warning Radar) and is primarily a missile attack early warning system. Two of these systems are located outside the u.s.: one in Greenland, the other in northern England at Fylingdales Moor. Their primary and highly sophisticated role is to detect and differentiate between the origins, targets and nature of approaching enemy missiles, but this level of capability is equally useful for tracking orbiting/decaying satellites or space debris.

The DOD also uses optical systems: the 'Ground-based Electro-optical Deep Space Surveillance' (GEDSS) set-up comprises three

The Air Force Maui Optical and Supercomputing (AMOS) site's telescopes used for tracking orbital debris.

telescopes spread across the globe, in New Mexico, Hawaii and the British Indian Ocean Territory of Diego Garcia. The GEDSS specifications and capabilities are impressive: it can 'see' objects in space as small as a basketball out to a distance of 20,000 km. Nevertheless, even this high degree of sensitivity has been improved upon as military planners have demanded ever more powerful tracking systems that would be able to scan vast swathes of sky in minute detail. A glossy promotional flyer

vapour within. With data accumulating over years and years, these developments enabled more substantial long-term studies of the world's climate.

WEATHER AND CLIMATE

One satellite – TRMM (Tropical Rainfall Measuring Mission), a Japan/USA project from 1997–2015 – returned the first substantive data on the anatomy of storms, the rainfall they deliver and the distribution of lightning bursts. An updated mission called GPM, for Global Precipitation Measurement, did just that, its Core Observatory satellite scanning rainfall over vast swathes of the Earth's surface with its microwave and radar instruments. GPM's data would be used as a

TRMM (Tropical Rainfall Measuring Mission) image sequence of a storm moving across the United States from the Pacific to the Atlantic, 2014.

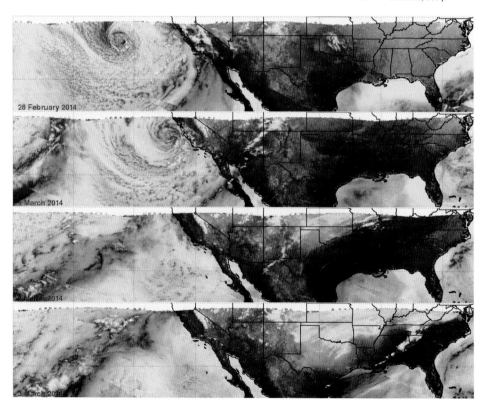

28 February 2014

1 March 2014

standard with which to calibrate further rainfall readings gathered by other satellites to provide a global set for climate studies. The GPM project was an international collaboration between the United States's NASA and Japan's JAXA (Japan Aerospace Exploration Agency), like TRMM, and illustrated a developing dependency on networks of satellites and scientists conducting Earth science with space-based systems. The early satellites used by scientists to study Earth, its atmosphere and oceans tended towards dedicated missions: a satellite scanning the atmosphere, another the oceans and so on. Later, single satellites were able to gather information on all manner of different environments. The European Space Agency's Envisat satellite bristled with nine scientific instruments, analysing atmospheric chemistry and pollution; sea surface temperatures (its AATSR – Advanced Along-Track Scanning Radiometer – acting as the global standard for sea surface temperature readings); the planet's surface and atmosphere's reflectance of sunlight; the heights of ocean waves and the distribution of sea ice. It provided continuity, too: its radiometer was a more powerful version of that built by the same team and flown on two preceding satellites, the first launched in 1991; by the end of Envisat's life these three instruments had provided scientists with twenty years of observation, adding greatly to the significance of the science carried out. Envisat was one of the largest satellites ever launched – the size of a bus and weighing in (on Earth) at well over 8 tonnes. In London we tried, at the time of the satellite's launch, to fit a full-sized configuration model of the Envisat into the Science Museum. It was too big to display in any gallery. Fortunately, with the opening of the Atmosphere gallery in 2010, an engineering model of its crucial AATSR was lofted above the displays, an eye in the sky on the heat generated below.

FORMATION FLYING

Other satellites can carry a smaller array of scientific tools but fly in formation, each member providing complementary sets of data for a

Vast blooms
of marine algae
– highly sensitive
to climate change –
captured by Envisat's
MERIS (Medium
Resolution Imaging
Spectrometer)
instrument off the
northwest coast
of Europe, 2011.

given period. The A-Train, shorthand for the 'Afternoon Constellation'
and a play on the Duke Ellington melody, comprised five satellites
from different nations all flying in sequence in approximately the same
polar orbit and passing over the equator within minutes of each other.
A-Train's CloudSat and CALIPSO satellites were launched on the same
rocket in 2006. CloudSat used radar to probe the interiors of clouds;
CALIPSO peered into man-made aerosols – tiny particles or droplets
lingering in the atmosphere – with a lidar (light detection and radar)

The Envisat satellite
being assembled at
the European Space
Agency's ESTEC
facility, Netherlands,
2002.

instrument, using laser light rather than radar's microwaves to aim at
objects. These satellites joined another two in orbit – Aqua, which had
been launched four years earlier, and PARASOL, two years after that.
Aqua was studying the Earth's water cycle with five instruments built

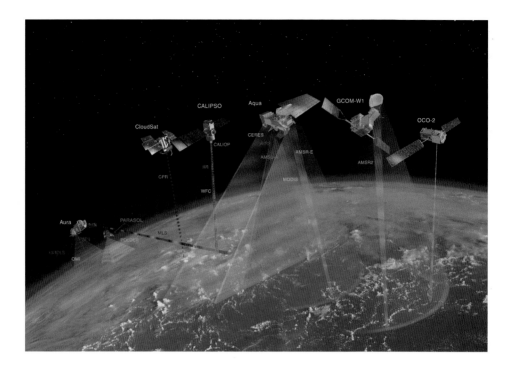

and operated by groups from the U.S., Japan and Brazil. It measured atmospheric humidity levels and temperatures, precipitation rates, ice, snow and soil moistures. PARASOL was using its own technique to measure atmospheric aerosols, while a fifth satellite, Aura, launched in 2004, used four instruments to study the Earth's ozone layer, and analyse air quality and the distribution of pollutant molecules. It took eight minutes for this A-Train of five satellites to pass overhead. CloudSat trailed Aqua by about 450 km (280 miles) in its orbit, and was some 90 km (56 miles) in front of CALIPSO. The satellites' ground track was repeated every sixteen days, or 233 orbits.

Artist's impression of the A-Train satellite formation with lines and shading depicting the area and reach of their instruments.

Another satellite constellation called GRACE (Gravity Recovery and Climate Experiment), comprising just two NASA satellites, in 2002 used a novel technique to accurately measure the Earth's gravity field. Better understandings of its distribution could then be factored in to studies on geological, oceanological and climatological phenomena.

The two satellites (nicknamed 'Tom' and 'Jerry') flew the same polar orbit about 220 km (137 miles) apart. They used a microwave ranging system to measure the exact distance between the two to an accuracy of 10 micrometres (that's about one-tenth the thickness of a human hair). When the leading satellite passed over a region of increased gravitational force, it was pulled slightly ahead in its motion – the distance between the two satellites increasing very slightly. When it had passed the anomaly, it would slow in response to the gravity dropping back in strength, before the second satellite passed over the same region reacting in a similar manner. By combining the minute changes in distance between the two satellites, as they were pulled and then released by the gravitational difference, with their respective positions at the time (measured by GPS satellites), a highly detailed map of the Earth's gravity field started to be built. With the make-up of Earth's gravity field dependent on the distribution of its mass, whatever form it takes, be it land, ice or water, the GRACE data generated was

Artist's impression of the two GRACE (Gravity Recovery and Climate Experiment) twin satellites orbiting Earth, *c.* 2002.

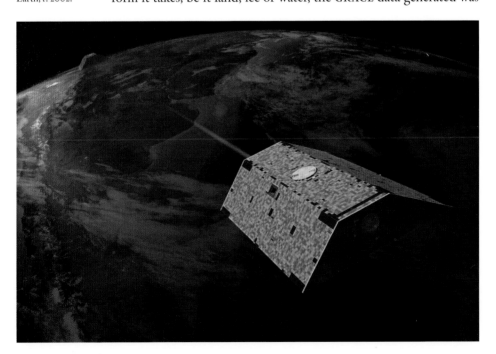

used to measure the shifting of water on the surface and deep in the
ocean, the storage of water on land and also its run-off, the movement
of water between ice sheets, glaciers and the ocean, and shifts in the
Earth's crust following earthquakes.

LOOKING OUT

In 1950 several top scientists had met in the U.S. to discuss the possibilities
for an International Geophysical Year (IGY) in which rockets could play
a crucial part, not only enabling improved investigation of the Earth's
upper atmosphere but initiating a new era of astronomy. Sounding
rockets would punch through the atmosphere to altitudes far beyond
those reached by instrument-carrying aircraft or balloons. Their view of
space would be unencumbered by the Earth's atmosphere, although the
detectors carried would have but a few minutes scanning the heavens
before plummeting back down. Satellites, though, could languish in
orbit and return long-term readings of all manner of cosmic bodies and
events. Like most of the terrestrially focused spacecraft, these would
sense electromagnetic radiation but now from across deep space. Theirs
would offer far clearer images of distant worlds and galaxies. They
would respond to forms of light quenched or stopped altogether by
the atmosphere. Infrared light (heat) would be pure and pristine, not
having had to compete with Earth's own maelstrom of infrared within
and below its atmosphere. For the first time, shorter wavelengths of light
and their sources – ultraviolet, X-ray and gamma ray – all of which are
blocked by the atmosphere, could be detected cleanly. From the 1960s
onwards a flotilla of space-based, orbiting astronomical observatories
sampling the broad range of electromagnetic radiation revolutionized
scientists' understanding of the universe: its composition, its age, even
the manner of its formation.

　　IRAS (Infrared Astronomical Satellite) and ISO (Infrared Space
Observatory), launched in 1983 and 1985 respectively, were able to see
through the clouds of dust that permeate huge regions of space and that

block the visible light emitted by stars. Even though these spacecraft were far clear of the Earth's atmosphere, their highly sensitive infrared detectors had to be shielded from stray heat (including the heat emanating from the rest of the satellite and its systems) by being chilled to approximately 2° Kelvin, which corresponds to −271° Celsius and is just two degrees above absolute zero, the coldest temperature that anything can be in the universe. Both the IRAS and ISO satellites used liquid helium to keep the detectors at this temperature. IRAS provided the first all-sky survey in the infrared, discovered some 350,000 infrared sources – many of them new galaxies – and identified discs of dust around stars, and several new comets including an unexpectedly long dust trail from comet Temple-2.

The Herschel space observatory, another infrared mission (but also carrying detectors sensitive to the adjacent sub-millimetre portion of the electromagnetic spectrum) launched several years after IRAS and ISO, used far more sensitive detectors, but these necessarily required even more protection from ambient infrared radiation. Herschel was therefore launched into an exotic orbit of the Sun called 'Sun–Earth Lagrangian 2' or L2. This location, named after the eighteenth-century astronomer and mathematician Joseph-Louis Lagrange, placed the

IRAS (Infrared Astronomical Satellite) infrared image of the entire sky with the bright band representing the plane of the Milky Way galaxy, 1983.

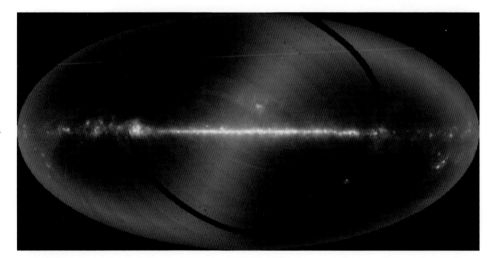

satellite over 1 million km out from Earth at the end of an imaginary straight line connecting the Sun, the Earth and Herschel. Normally, a spacecraft (or planet) orbiting the Sun at a greater distance than Earth would process around the Sun more slowly than Earth – its 'higher' orbit dictates that it would orbit more slowly, the velocity needed to balance the weaker pull of gravity of the Sun at this distance proportionately lower. But at Sun–Earth L2, where the Sun and Earth are in line, Earth's pull of gravity on the spacecraft must be added to that of the Sun's. The sum effect on the spacecraft is as if it were orbiting a larger Sun with an accordingly larger pull of gravity, thus requiring the spacecraft to travel at a slightly increased velocity in order to remain in that orbit. At Sun–Earth L2, the net gravitational effect on the spacecraft allows it to orbit the Sun at the same rate as the Earth does, and so the three bodies remain in line. In reality, the Sun–Earth L2 point is not fixed because of the perturbations in Earth's motion, and the spacecraft will appear to trace out a loop at this point as observed from Earth.

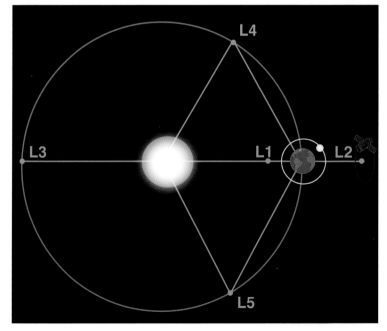

The five points, L1–L5, where a spacecraft orbiting the Sun can do so in a constant pattern in relation to both the Sun and the Earth. L5 is used by spacecraft that need to be shielded from the heat of the Sun, Earth and Moon.

With Herschel thus placed and orientated so that its detectors were always pointing away from the Sun and the Earth, it would have an unobstructed view of the heavens ensuring also that no stray light and heat from the Sun, Earth or Moon disturbed the detectors' sensitive readings. Herschel confirmed the presence of oxygen molecules in deep space, and suggested also that Earth's water may have originated from comets and that water vapour was present on Ceres, the largest of the asteroids. This added to the growing evidence that water, once thought to be rare or even absent in space, is in fact widespread.

VIOLENT UNIVERSE

The types of satellite observatories mentioned capturing infrared or sub-millimetre light were focused on relatively quiet, placid and cool targets in space – dust clouds, comets and asteroids. Others were trained on the most violent and energetic bodies and phenomena in space and had therefore to be sensitive to the high-energy electromagnetic radiation – X-rays and gamma rays – that these targets emit. The ROSAT (abbreviated from Röntgen Satellite named after the discoverer of X-rays, Wilhelm Röntgen) showed all bodies in space emit X-rays but that the most intense were thrown out during cosmic death-throes – when galaxies collide or when stars are swallowed by black holes. Other X-ray satellites, including NASA's Chandra, the ESA's XMM-Newton and Spitzer, added to the observations, while one X-ray observatory – the Spacelab-2 X-ray telescope carried aboard the *Challenger* Shuttle on what turned out to be its penultimate mission before its tragic loss in 1986 – used an innovative detector to image high-energy X-rays (which would pass straight through conventional detectors). It was positioned in *Challenger*'s open payload bay while orbiting Earth and produced the first high-energy X-ray map of the centre of our own galaxy, the Milky Way.

Another satellite observatory – Swift – tracked the most explosive events known in the universe: gamma ray bursts or GRBs. These

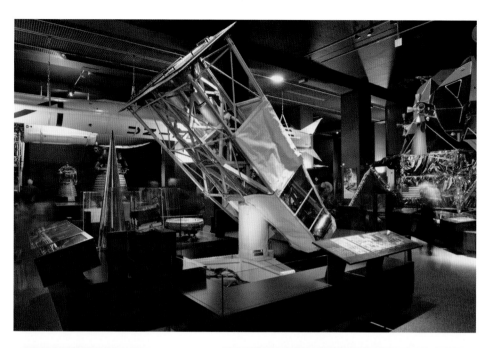

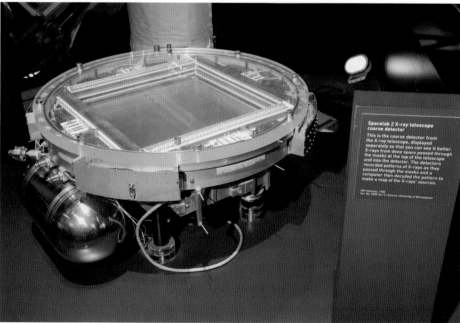

Spacelab 2 X-ray telescope coarse detector

This is the coarse detector from the X-ray telescope, displayed separately so that you can see it better. X-rays from deep space passed through the masks at the top of the telescope and into the detector. The detectors recorded patterns of X-rays as they passed through the masks and a computer then decoded the pattern to make a map of the X-rays' sources.

XPT detector, 1980
Inv. No. 2005-32 / 3 | Source: University of Birmingham

The Spacelab-2 X-ray telescope display in the Exploring Space gallery of the Science Museum, 2007.

The Spacelab-2 X-ray telescope's coarse photon detector in the Exploring Space gallery of the Science Museum, 2007.

occur when stars of a certain size, at the point their nuclear fuel is depleted, collapse in on themselves, becoming increasingly dense and massive. Some will become so dense that even light cannot escape their gravitational clutches and they transform into black holes. But others will explode as supernovae, leaving a remnant composed entirely of neutrons. These neutron stars are the densest and smallest stars in the universe – they can be as little as a few kilometres in diameter yet weigh as much as two of our galaxy's Suns. Scientists using satellites like Swift concluded that GRBs are generated during the supernovae of exploding stars or when neutron stars collide. 'Swift' is so called because of its rapid response to GRBs – as soon as one is detected it

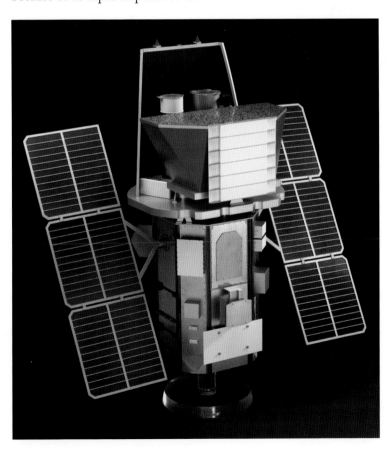

Scale model (1:10) of Swift gamma ray space observatory, c. 2002.

homes in on its origin so that its other detectors can catch some of the afterglow radiated at different wavelengths. The positions of these bursts are relayed immediately to radio telescopes on Earth that can then be swivelled towards the GRB's source and so register the radiation being emitted at radio wavelengths.

OUR PLACE IN THE UNIVERSE

GRBs come from all directions indicating origins from way beyond our own galaxy and long into a distant time – the further out into the universe you look, the older the objects and phenomena you are looking at. One satellite, however, in 2004 looked towards Earth's own effect on the universe. Gravity Probe B carried gyroscopes and a telescope. The telescope, fixed to the main structure of the satellite, located a chosen star, IM Pegasi (HR 8703), and the satellite's orientation was programmed so that it would point continuously towards that star. The gyroscopes would be spun up and aligned so that their axes of spin pointed towards the star. Gyroscopes remain orientated in the same direction as long as they continue to spin, so these would continue to point to the star. Once this was set up, the scientists waited. They were testing and measuring Einstein's general theory of relativity and specifically its prediction that the Earth distorts the part of the universe it is situated in – that it actually, very slightly, warps spacetime (the fabric of the universe) – and this effect would be registered with the slow drift of the gyroscopes' axes of spin. The drift would be observable as the gyroscopes would no longer be pointing exactly at the reference star. The gyroscopes would measure a second predicted effect of general relativity: the Earth's dragging of spacetime as it spins. This is rather akin to a toy ball spinning in a bathtub of soapsuds, which pulls some of the bubbles around with it as it spins.

The satellite operated for one year, and the results confirmed and measured the effects predicted by general relativity to an unprecedented degree. The mission team collected a cabinet full of trophies for what

The Gravity Probe-B
satellite undergoing
acoustic tests prior
to launch, *c.* 2003.

One of Gravity
Probe-B's four
gyroscopic spheres,
c. 2003.

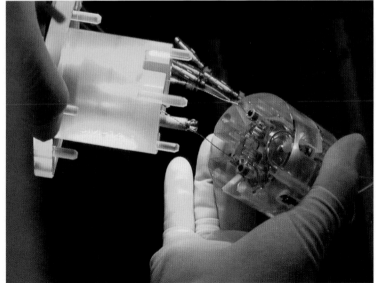

it had achieved. It was one of the most sophisticated spacecraft ever launched; the gyroscope spheres themselves, made out of pure quartz, were the most spherical objects ever made, meaning, in the team's words, 'that every point on the surface of the gyroscope [sphere] is the exact same difference from the centre of the gyroscope to within 0.0000003 inches'.

THE BEGINNING

Other science satellites probed the origins of the universe. The COBE satellite (Cosmic Background Explorer) was launched to examine further the radiation remnants of the Big Bang, the principal scientific theory describing the beginning of the universe. That such primordial radiation should exist was predicted by the Soviet scientists A. G. Doroshkevich and Igor Novikov back in 1964 and was then discovered by Arno Penzias and Robert Wilson at Bell Laboratories in the U.S. They were working on improved microwave receivers for radio astronomy and were using a horn antenna to collect test radio waves reflected from the giant Echo balloon satellites. The signals were so faint they needed to eliminate all extraneous radio 'noise'. Try as they might they could not eliminate a mysterious background hum that appeared to be coming from all directions in space. They even cleaned out the pigeon nests and droppings scattered within the horn in case they were the cause of the unwanted sound. Other space scientists were actively searching for cosmic background radiation, and when Penzias and Wilson were by chance talking with astrophysicists Robert Dicke, Jim Peebles and David Wilkinson at Princeton University, they realized that this was the source of the constant radio noise they were picking up from their antenna. COBE mapped this residual radiation from the Big Bang showing also its faint ripples, the beginnings of density fluctuations that would lead eventually to the dappled universe we know, where galaxies cluster amid vast tracts of empty space. COBE's colourful map of the radiation made the front pages, and a follow-up mission – WMAP (Wilkinson Microwave Anisotropy Probe) – added yet more detail to scientists' understandings of the beginning of everything. But there was another science satellite that made regular headlines, from the build-up to its launch in the Space Shuttle, the problems it encountered, the hazardous space walks astronauts performed to fix and service it and the astonishing pictures it returned of the universe. This was the Hubble Space Telescope.

SPECIAL HUBBLE

The Hubble Telescope, honouring the astronomer Edwin Hubble, who provided the key evidence of a universe expanding from an initial big bang, had been planned for many years: a large orbiting telescope that would scan the heavens free from Earth's obscuring atmosphere. It was launched by a Shuttle in 1990, but almost immediately showed a severe problem with its optical systems. Its primary mirror, which bounced the incoming light from stars and galaxies onto the telescope's detectors, was flawed – minutely so, yet sufficient to give Hubble blurred vision. The telescope had been designed for regular

Shades of blue and purple in this image depict temperature variations in the early stages of the universe. Generated from data gathered by the COBE spacecraft, 1992.

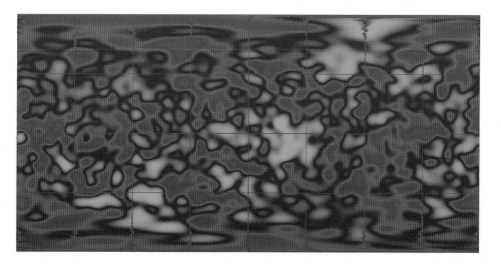

Beach ball with a
map of the cosmic
background radiation
as measured by WMAP,
the successor satellite
to COBE, 2003–11.

servicing by astronauts, and the first such mission installed a corrective
device to nullify the aberration. The flight was successful, and Hubble
started to return crystal-clear images as originally intended. There were
four more servicing missions with astronauts replacing worn-out or
redundant instruments and components during marathon space walks.
The mission had become a unique blend of science and spectacle,
where the most sophisticated science was enabled by the raw courage
of the astronauts. And yet it was the images of space that put Hubble
into a category of its own as a satellite mission; the name resonates
with millions – like Apollo, Armstrong or Gagarin. The pictures
spoke to the investigating scientists but also to everyone else. Their
ethereal beauty transcended science and art to say something of the

majesty and mystery of the universe. Its scale. Its depth. Its distance. The images spoke of our place in the universe, reassuring us that we were probing new heights yet reminding us there is so much we still do not know and understand. And Hubble revealed more worlds in other star systems. Decades earlier the only planets known were those of the solar system. Thousands of exoplanets orbiting other stars have now been discovered, but these are only a tiny fraction of the number estimated to exist.

Galileo Galilei had shown the Moon to be gnarled and battered with craters and mountains: a very real world; a destination that might perhaps be reached; nothing mythical or sublime about this heavenly

The Hubble Space Telescope, seen from the *Atlantis*.

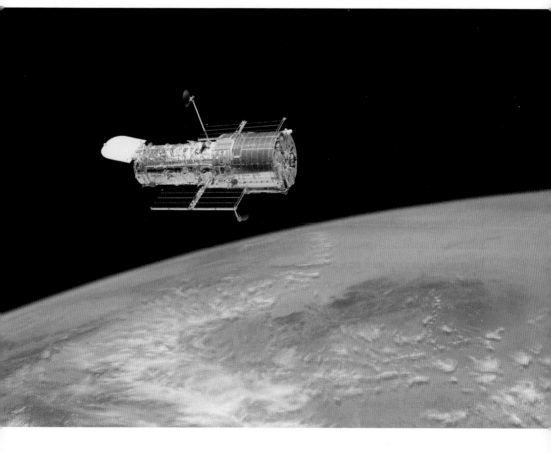

A Hubble Space
Telescope infrared
image of the gas and
dust of the Horsehead
nebula, 2013.

orb. He discovered the four large moons of Jupiter and showed
them to be orbiting the giant planet as Earth does the Sun. In 1989
NASA launched an artificial satellite to the Jovian system in Galileo's
honour. The eponymous spacecraft orbited the gas giant in vast loops,
returning images and data of all the large moons and showing them to
be extraordinary worlds – one a sulphurous, volcanic realm; the other
three with copious quantities of frozen or liquid water.

The Cassini mission to Saturn acknowledged the seventeenth-
century astronomer and discoverer of a gap – the Cassini Division
– in the rings of Saturn. It too performed huge sweeping orbits of its
host planet, tracing out the petals of a cosmic flower, and turned its

sensors onto two of Saturn's large moons, Enceladus and Titan. Cassini discovered icy plumes bursting out of Enceladus and, along with its deployed Huygens probe (Christiaan Huygens was a seventeenth-century Dutch astronomer and discoverer of Titan), showed Titan to be a strange, Earth-like world but with rain, rivers and oceans of liquid hydrocarbons. Its chemistry was thought to resemble that of pre-biotic Earth – similar to that to be found on our own planet prior to the appearance of life.

Venus, Earth's sister planet in size, had always been coy, its blankets of cloud obscuring what lay beneath. Pluralists presumed it, like Mars, no doubt harboured life: according to Emanuel Swedenborg in his *The Earth in our Solar System* of 1758, 'If Venus were without population, then the Earth must be similarly lacking, and reciprocally, if the Earth

Io, one of Jupiter's Galilean moons, imaged by the Galileo orbiter, 1996.

were populated, Venus must be populated too.' Gradually, twentieth-century observation showed our solar system twin to be a far-from-hospitable environment for terrestrial-type life. Spacecraft flew by and even landed. The Magellan satellite orbiter penetrated the planet's dense clouds to reveal a rocky surface with few impact craters. The views suggested recent (in geological time) volcanic activity where molten rock spewed and renewed the planet's surface, obliterating immediate traces of the impacts that peppered the planets in the early times of the solar system.

What of Mars? The planet that had fired the imaginations of H. G. Wells, Konstantin Tsiolkovsky, Fridrikh Tsander, Sergei Korolev and so many others. Mars tantalized with its menacing red glow; perhaps life existed there, perhaps not. It drew attention with its mystery; a destination to reach and understand. The first space-age close-ups of the planet disappointed those hoping to see vegetation, perhaps even bodies of water. Mariner 9 was the first to orbit another planet besides Earth, and returned pictures of a Mars with barren, cratered landscapes, little different to our Moon. There were enormous canyons and gullies, mountains and plains but no evidence of plant let alone animal life. Since then space agencies have sent a flotilla of spacecraft, including five orbiting satellites, to Mars. Their sensors and cameras have returned ever-improving images of the planet. We know more of Mars now than we do of Earth's deepest oceans. We know it was once wet with oceans and running water.

We know also that life on Earth is far more resilient than we ever once thought, with microbes thriving in the hottest, coldest, most acidic or alkaline environments thought possible. Some even live deep underground in rock, using iron as a source of energy. Might Mars, therefore, have once harboured primitive life? Might it still do so beneath its freezing, radiation-dowsed surface?

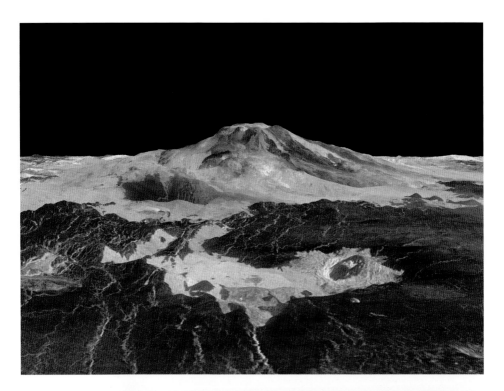

I138 Computer-
generated image of the
surface of Venus from
data returned by the
Magellan spacecraft,
1991. The 2-km-high
Sif Mons volcano is
clearly visible.

Mariner 9 image of
Promethei Sinus,
Mars, 1972. The frame
is 450 km from top to
bottom.

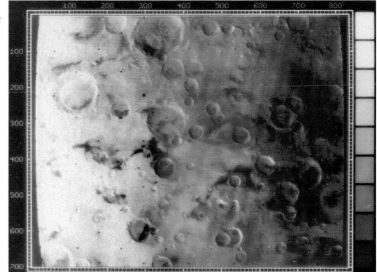

ROBOT OR HUMAN

More scientific missions will be launched to Mars to orbit, to land and to rove. The data returned will continue to mount. Among them may be an answer to whether life has ever existed on the planet. And yet many scientists feel that despite the increasing sophistication of robot spacecraft, it will need a human touch to provide the ultimate answer. Astronauts and cosmonauts can think on their feet and improvise in ways the satellites, landers and rovers cannot. Sending humans to Mars will remain immensely difficult. The billions of dollars spent on going to the Moon will pale in comparison to that needed to (first work out how to) land a human being on Mars. The immense Apollo project would rank as a gentle paddle in the waters of Earth's shoreline compared to a trans-oceanic swim of going to Mars. But still, Mars pulls at humanity's inquisitiveness. It continues to fire the imaginations of scientists, engineers and enthusiasts just as it did back in the late nineteenth and early twentieth centuries. Those dreamers and practitioners propelled us into Earth's orbit and beyond. They launched the age of the satellite and of space. The flocks of satellites

A view from the 'Kimberley' formation on Mars taken by NASA's Curiosity rover. The colours are adjusted so that rocks look approximately as they would if they were on Earth, to help geologists interpret the rocks. This 'white balancing' to adjust for the lighting on Mars overly compensates for the absence of blue on Mars, making the sky appear light blue and sometimes giving dark, black rocks a blue cast, 2014.

orbiting silently and almost invisibly high above us now help and tell us more of our own planet. There is today, in effect, a domain of Earth that sits above the planet's geosphere and biosphere and the layers of its atmosphere. It comprises strata and constellations of artificial satellites.

Konstantin Tsiolkovsky, his contemporaries and those influenced by him, imagined the exploration of space, reaching the planets and beyond. Putting satellites and space stations in orbits of Earth were but a beginning for travelling still further out into the universe. It was our destiny to do so. Such spiritual guidance has all but disappeared today, and yet for some there remains a perceived *necessity* to reach out beyond our home planet. The vision is one of survival. If humanity is to prosper, then Earth's finite supply of natural resources will one day need supplementing from other worlds. Our future depends on our having routine access to the solar system which, despite all that has been achieved in the space age, we do not yet have.

Getting into orbit and beyond remains tremendously expensive. Every rocket launch costs many tens of millions of dollars, and the vehicle is used just the once. British space engineer Alan Bond likens it to crossing the Atlantic in a jet airliner which is thrown away at the end of the flight. Each satellite, each space probe, each astronaut crew ever launched since Sputnik in 1957 has used a rocket system Tsiolkovsky would likely recognize; a vehicle constructed of stages, each a rocket in its own right, that fall away to their destruction in sequence on the way to orbit – effective but hugely wasteful and inefficient.

The high cost of launches impacts the quality of satellites and spacecraft launched; they have to be engineered to the highest possible standard to ensure they work; repair and replacement once in space is either difficult or impossible. Bond pictures instead a new space age where reaching orbit and beyond becomes as easy as flying the Atlantic. Rockets would be reused, just as aeroplanes are, and the cost of launching satellites would plummet. For decades he has worked on designing such a satellite launch vehicle – one where on-board propellant would constitute far less of the vehicle's total mass, so dispensing with the weight-saving arrangement of separate stages. This single-stage-to-orbit vehicle would use a novel rocket engine capable of breathing atmospheric oxygen before switching over to the much smaller supply of stored onboard oxygen for the final part of the ascent to orbit by conventional rocket propulsion.

The artificial satellite was launched into space with the type of technology developed during the Second World War. It followed, seemingly delivering on, the dreams and imaginations of space exploration from many years earlier. Many of those visions remain, however. Alan Bond, like many others who have worked in the space sector, was weaned on the science fiction of the 1950s – on the radio, in the cinema and in magazines. In those tales of daring we flew into orbit and on through the solar system as a matter of course. New adventures were to be had in every episode. Such will be the case when reaching orbit is routine and cheap. Satellites will then become truly commonplace, and a new space age will be born.

TIMELINE

1611 Earliest celestial application of the word 'satellite' when Johannes
      Kepler refers to each of the moons of Jupiter recently discovered by
      Galileo

1869 Edward Hale one of the first to suggest a use for a satellite in his story
      'The Brick Moon'

1929 Herman Potočnik describes in his book *Das Problem der Befahrung
      des Weltraums – der Raketen-Motor* (The Problem of Space Travel: The
      Rocket Motor) how a space station orbiting the Earth above the equator
      at an altitude of 36,000 km would appear stationary in the sky

1929 Hermann Oberth describes space stations with spherical living quarters
      in his book *The Ways to Spaceflight*

1942 The first successful test launch of an A-4 rocket carried out by the
      German Army at its Peenemünde research centre on the Baltic coast

1945 Arthur C. Clarke describes a working geostationary satellite television
      broadcasting system in the *Wireless World* article 'Extra-terrestrial
      Relays'

1946 RAND Corporation publishes 'Preliminary Design of an Experimental
      World-circling Spaceship'

1951 International Astronautical Federation's second congress meets in
      London with the artificial satellite as its theme

1954 Politburo instructs the Soviet Union's main missile design bureau
      (OKB-1) to start production of the R-7, which would become the
      world's first intercontinental ballistic missile

1954   Sergei Korolev submits Tikhonravov's 'On the Artificial Satellite of the Earth' report to the Politburo

1955   Academy of Sciences creates first Soviet organization devoted to space flight

1956   U.S. Air Force authorizes WS-117 reconnaissance satellite programme

1957–8 International Geophysical Year

1957   Sputnik, first artificial satellite of the Earth, launched by the Soviet Union

1957   Laika in USSR's *Sputnik* 3 becomes the first animal to orbit the Earth

1958   U.S. establishes the Advanced Research Projects Agency (ARPA) in response to Sputnik

1958   U.S. launches its first satellite, Explorer

1958   U.S. establishes the National Aeronautics and Space Administration (NASA) in response to the Soviet space challenge

1959   U.S. launches first Transit navigation satellite

1960   U.S. launches the Echo 1 passive communications satellite

1960   U.S. launches first meteorological satellite, TIROS

1960   U.S. Discoverer XIV reconnaissance satellite returns film showing Soviet missile numbers to be far fewer than believed

1960   USSR dogs Belka and Strelka become the first creatures to return safely from orbit

1961   Yuri Gagarin becomes the first human to orbit the Earth

1962   U.S. Communications Satellite Act underpins the commercialization of satellite communications

1962   U.S. Telstar satellite relayed the first live transatlantic television pictures

1964   The International Telecommunications Satellite Organization (subsequently known as Intelsat) formed

1964   U.S. Syncom 3 becomes the first geostationary (communications) satellite

1965   U.S. 'Early Bird' communications satellite (Intelsat-1) launched

1966   USSR Luna 10, the first spacecraft to orbit the Moon

1967   'Our World', the first live global satellite television programme

1968   U.S. *Apollo 8*, the first spacecraft with human crew to orbit the Moon

1970   U.S. launches the first X-ray astronomy satellite, *Uhuru*

1971   USSR launches world's first space station, Salyut

1971   U.S. Mariner 9, an unmanned space probe, becomes the first artificial satellite of another planet (Mars)

1972   U.S. launches first Earth-observation satellite, Landsat (originally Earth Resources Technology Satellite)

1975   USSR Venera 9, the first spacecraft to orbit Venus

1978   U.S. launches first experimental Navstar Global Positioning Satellite

1983   International consortium launches first Infrared Astronomy Satellite, IRAS

1989   U.S. launches the Cosmic Background Explorer (COBE) cosmological satellite

1989   U.S. Galileo spacecraft the first to orbit Jupiter

1990   International consortium launches the Hubble Space Telescope into Earth's orbit

1997   U.S. launches the first Iridium communications satellite

2000   U.S. Near Earth Asteroid Rendezvous – Shoemaker (NEAR Shoemaker) spacecraft is the first to orbit an asteroid

2002   European Space Agency (ESA) launches its largest environmental monitoring satellite, Envisat

2004   International Cassini spacecraft becomes the first to orbit Saturn

2004   International Swift satellite, the first dedicated gamma ray burst (GRB) mission

2014   Rosetta mission spacecraft becomes the first to orbit a comet

SELECT BIBLIOGRAPHY

Alper, Joel, and Joseph P. Pelton, eds, 'The Intelsat Global Satellite System',
    in *Progress in Astronautics*, ed. Martin Summerfield, XCIII (New York,
    1984)

Benjamin, Marina, *Rocket Dreams: How the Space Age Shaped our Vision
    of a World Beyond . . .* (London, 2003)

Butrica, Andrew J., ed., *Beyond the Ionosphere: Fifty Years of Satellite
    Communication* (Washington, DC, 1997)

Day, Dwayne A., 'Cover Stories and Hidden Agendas', in *Reconsidering
    Sputnik: Forty Years since the Soviet Satellite*, ed. Roger D. Launius et al.
    (Amsterdam, 2000)

Dunér, David, 'Venusians: The Planet Venus in the 18th-century Extra-
    terrestrial Life Debate', *Journal of Astronomical Data*, XIX/1 (2013)

Geppert, Alexander C. T., ed., *Imagining Outer Space: European
    Astroculture in the Twentieth Century* (Basingstoke, 2012)

King-Hele, Desmond, *A Tapestry of Orbits* (Cambridge, 1992)

Krige, J., and A. Russo, *A History of the European Space Agency, 1958–1987*,
    vol. I: *The Story of ESRO and ELDO* (Noordwijk, 2000)

—, and L. Sebesta, *A History of the European Space Agency, 1958–1987*, vol. II:
    The Story of ESA (Noordwijk, 2000)

Krige, John, Angelina Long Callahan and Ashok Maharaj, *NASA in the
    World: Fifty Years of Collaboration in Space* (New York, 2013)

McCray, W. Patrick, *Keep Watching the Skies: The Story of Operation
    Moonwatch and the Dawn of the Space Age* (Princeton, NJ, 2008)

Mack, Pamela E., *Viewing the Earth: The Social Construction of the Landsat Satellite System* (Cambridge, 1990)

Millard, Doug, ed., *Cosmonauts: Birth of the Space Age* (London, 2014)

Neufeld, Michael J., *Von Braun: Dreamer of Space, Engineer of War* (New York, 2007)

Ordway III, Frederick I., and Randy Liebermann, eds, *Blueprint for Space: Science Fiction to Science Fact* (Washington, DC, and London, 1992)

Parkinson, Bob, ed., *Interplanetary: A History of the British Interplanetary Society* (London, 2008)

Parks, Lisa, *Cultures in Orbit: Satellites and the Televisual* (Durham, NC, and London, 2005)

—, and James Schwoch, eds, *Down to Earth: Satellite Technologies, Industries and Cultures* (New Brunswick, NJ, and London, 2012)

Sachdev, D. K., ed., *Success Stories in Satellite Systems* (Reston, VA, 2009)

Shepherd, L. R., 'The Artificial Satellite: An Introduction to the Symposium on Satellite Vehicles at the Second International Congress on Astronautics, London, 1951', *Journal of the British Interplanetary Society*, X/6 (November 1951)

Siddiqi, Asif A., 'Korolev, Sputnik and the IGY', in *Reconsidering Sputnik: Forty Years since the Soviet Satellite*, ed. Roger D. Launius et al. (Amsterdam, 2000)

Von Braun, Wernher, 'The Importance of Satellite Vehicles in Interplanetary Flight', *Journal of the British Interplanetary Society*, X/6 (November 1951), pp. 237–44

Winter, Frank H., *Prelude to the Space Age: The Rocket Societies, 1924–1940* (Washington, DC, 1983)

OTHER

Gravity Probe B Experiment, 'Testing Einstein's Universe', Press Kit, April 2004, National Aeronautics and Space Administration, p. 50; http://einstein.stanford.edu/content/press-media/press_kit_2004, accessed 17 February 2015

Parkinson, Bradford, 'GPS for Humanity', http://scpd.stanford.edu,
     2 May 2012
'Preliminary Design of an Experimental World-circling Spaceship', Report
     no. SM-11827, Contract W33-038 ac-14105, Douglas Aircraft Company,
     Inc., Santa Monica Plant Engineering Division, CA, 2 May 1946,
     p. VI; www.rand.org, accessed 13 January 2015; the RAND (Research and
     Development) think tank was formed 1 October 1945 by the Douglas
     Aircraft Company under contract from the U.S. Government

ACKNOWLEDGEMENTS

Many people have helped in the delivery of this book, but I should like to thank, in particular, Peter Morris for his thoughts on the early iterations; David Exton and John Herrick for helping trawl the Science Museum photographic archives, and especially for locating the image from the 1957 International Geophysical year exhibition; Jennie Hills for the new photography; Jasmine Rodgers and her team in the Science and Society Picture Library for their patience in dealing with endless questions about image resolution and file size; Becky Storr and colleagues at Blythe House for providing a haven of peace and quiet away from the bustle of the Museum offices; Suzann Parry and the rest of the British Interplanetary Society staff for making my visits to the Society's library and archives such a pleasure; Vivian Constantinopoulos at Reaktion Books for her expert comments on the manuscript, together with Amy Salter and Becca Wright for dealing so calmly with the blizzard of text, tables, images and credits I blew in their direction; Dan Mercer at Iridium who put me in touch with his transatlantic colleagues; and Stephanie, whose image-hunting skills kept me on track when too many deadlines converged.

PHOTO ACKNOWLEDGEMENTS

The author and publishers wish to express their thanks to the following for illustrative material and/or permission to reproduce it.

Courtesy Heriberto Arribas Abato: p. 59; © *Daily Herald* Archive/ National Media Museum/Science & Society Picture Library: pp. 35, 61 (top), 110 (bottom); European Space Agency: pp. 130, 165; European Space Agency/A. Van Der Geest: p. 164; Infrared Processing and Analysis Center, Caltech/JPL: p. 169; Iridium Communications Inc.: p. 120 (top and bottom); EADS Astrium/Science Museum, London: p. 113; © *Manchester Daily Express*/Science Museum/Science & Society Picture Library: p. 77; Don S. Montgomery, USN (Ret.)/United States Department of Defense: p. 100; NASA: pp. 29, 36, 37 (bottom), 38, 50, 70, 71, 72, 83, 87, 88, 96, 97, 98 (top and bottom), 101, 102, 105, 115, 133, 138, 140, 150, 162, 166, 175 (top and bottom), 179, 183 (bottom); NASA/DMR/COBE Science Team: p. 177; NASA/ESA/Hubble Heritage Team: p. 180; NASA/Frederick W. Kent Collection, University of Iowa Archives: p. 52; NASA/Bill Hrybyk: pp. 12–13; NASA/JPL: pp. 61 (top), 66, 183 (top); NASA/JPL-Caltech: pp. 53, 73, 152, 167; NASA/JPL-Caltech/MSSS: p. 184; NASA/JPL/University of Arizona: p. 181; NASA/Crew of STS-132: p. 99; NASA/Marshall Space Flight Center Collection: pp. 74, 143; NASA/Naval Research Laboratory: p. 64 (top); NASA Orbital Debris Program Office: p. 159; © NASA/Science & Society Picture Library: pp. 54, 62, 89, 104 (top), 116, 123; NASA/A. Siddiqi: p. 61 (bottom); courtesy the National Reconnaissance Office (Center

INDEX